Symmetry Problems.
The Navier–Stokes Problem.

Alexander G. Ramm

Synthesis Lectures on Mathematics and Statistics

Editor
Steven G. Krantz, *Washington University, St. Louis*

Symmetry Problems. The Navier–Stokes Problem.
Alexander G. Ramm
2019

PDE Models for Atherosclerosis Computer Implementation in R
William E. Schiesser
2018

An Introduction to Partial Differential Equations
Daniel J. Arrigo
2017

Numerical Integration of Space Fractional Partial Differential Equations: Vol 2 –
Applicatons from Classical Integer PDEs
Younes Salehi and William E. Schiesser
2017

Numerical Integration of Space Fractional Partial Differential Equations: Vol 1 –
Introduction to Algorithms and Computer Coding in R
Younes Salehi and William E. Schiesser
2017

Aspects of Differential Geometry III
Esteban Calviño-Louzao, Eduardo García-Río, Peter Gilkey, JeongHyeong Park, and Ramón
Vázquez-Lorenzo
2017

The Fundamentals of Analysis for Talented Freshmen
Peter M. Luthy, Guido L. Weiss, and Steven S. Xiao
2016

Symmetry Problems.   he Navier–Stokes Problem.
Alexander G. Ramm

ISBN: 978-3-031-01287-7   paperback
ISBN: 978-3-031-02415-3   ebook
ISBN: 978-3-031-00261-8   hardcover

DOI 10.1007/978-3-031-02415-3

A Publication in the Springer series
SYNTHESIS LECTURES ON MATHEMATICS AND STATISTICS
Lecture #23
Series Editor: Steven G. Krantz, *Washington University, St. Louis*
Series ISSN
Print 1938-1743 Electronic 1938-1751

Symmetry Problems.
The Navier–Stokes Problem.

Alexander G. Ramm
ramm@ksu.edu

SYNTHESIS LECTURES ON MATHEMATICS AND STATISTICS #23

ABSTRACT

This book gives a necessary and sufficient condition in terms of the scattering amplitude for a scatterer to be spherically symmetric. By a scatterer we mean a potential or an obstacle. It also gives necessary and sufficient conditions for a domain to be a ball if an overdetermined boundary problem for the Helmholtz equation in this domain is solvable. This includes a proof of Schiffer's conjecture, the solution to the Pompeiu problem, and other symmetry problems for partial differential equations. It goes on to study some other symmetry problems related to the potential theory. Among these is the problem of "invisible obstacles." In Chapter 5, it provides a solution to the Navier–Stokes problem in $\mathbb{R}^3$. The author proves that this problem has a unique global solution if the data are smooth and decaying sufficiently fast. A new *a priori* estimate of the solution to the Navier–Stokes problem is also included. Finally, it delivers a solution to inverse problem of the potential theory without the standard assumptions about star-shapeness of the homogeneous bodies.

KEYWORDS

Helmholtz equation, symmetry problems, Navier–Stokes problem, Inverse problem of potential theory

To Luba

Contents

Preface

The contents of this book is taken from the author's published papers cited in the bibliography. The goals of this work are:

1. To give a necessary and sufficient condition in terms of the scattering amplitude for a scatterer to be spherically symmetric. By a scatterer we mean a potential or an obstacle.

2. To give necessary and sufficient conditions for a domain to be a ball if an over-determined boundary problem for the Helmholtz equation in this domain is solvable.

   This includes a proof of Schiffer's conjecture, the solution to the Pompeiu problem, and other symmetry problems for partial differential equations.

3. To study some other symmetry problems related to the potential theory. Among these is the problem of "invisible obstacles" [29].

4. To give a solution to the Navier–Stokes problem in $\mathbb{R}^3$. The author proves that this problem has a unique global solution, dependent continuously on the data, if the data are smooth and decaying sufficiently fast.

   A new *a priori* estimate of the solution to the Navier–Stokes problem is given.

   The absence of turbulent behavior of the fluid in the whole space is proved under the assumption that the data are smooth and rapidly decaying.

   This problem is one of the millennium problems.

5. To give a solution to inverse problem of the potential theory without the standard assumptions about star-shapeness of the homogeneous bodies.

   The author did not discuss in this book his theory for creating materials with a desired refraction coefficient. This theory is developed in the monographs [37], [38], and in his many papers mentioned in these monographs. This theory is based on the wave-scattering theory by many small impedance particles the author has developed. The importance of practical solution of the problem of producing one such particle with a prescribed boundary impedance is emphasized in [38] and [40].

   The author's theory of inverse scattering problems with non-over-determined scattering data is not discussed. This theory is developed in the monograph [49] and papers cited therein, in particular [45], [46], and [54].

   In this small book the results of the author are presented.

The author presents the basic ideas and results in a self-contained way.

The reader can study each chapter independently of the others.

The author thanks the Journals where his papers were published and used in writing this book.

Alexander G. Ramm
February 2019

CHAPTER 1

Introduction

The goal of this introduction is to briefly describe main results presented in this book.

In Chapter 2, we consider the scattering problem by a potential or an obstacle (see [49]).

The scattering problem by a potential consists of finding the scattering solution $u(x, \alpha, k)$ from the Schrödinger equation

$$\left[\nabla^2 + k^2 - q(x)\right]u = 0 \quad \text{in} \quad \mathbb{R}^3, \tag{1.1}$$

$$u = u_0 + v, \quad u_0 = e^{ik\alpha \cdot x}, \tag{1.2}$$

$$v_r - ikv = o\left(r^{-1}\right), \quad r \to \infty. \tag{1.3}$$

Here $r := |x|$, $k > 0$ is a constant, $\alpha \in S^2$ is the direction of the incident plane wave, S^2 is the unit sphere in $\mathbb{R}^3$, v is the scattered field satisfying the radiation condition (1.3), $\alpha \cdot x$ is the dot product of two vectors, $v_r = \frac{\partial v}{\partial r}$. The physical meaning of k^2 is the energy of a particle.

One can prove that

$$v = A(\beta, \alpha, k)\frac{e^{ikr}}{r} + o\left(r^{-1}\right), \quad r \to \infty, \quad \beta = \frac{x}{r}. \tag{1.4}$$

The function $A(\beta, \alpha, k)$ is called the scattering amplitude, β is the unit vector in the direction of the scattered field. The potential

$$q \in Q_b := \left\{q : q = \bar{q}, q \in L^2(B_b), q = 0 \text{ if } |x| > b\right\}, \quad B_b := \left\{x : x \in \mathbb{R}^3, |x| \le b\right\}.$$

The problem solved in Chapter 2 consists of finding necessary and sufficient conditions on the scattering amplitude $A(\beta, \alpha, k)$ for the potential q to be spherically symmetric, $q(x) = q(r)$.

A similar problem is studied for scattering by obstacles; see [49]. By an obstacle, a bounded domain in $\mathbb{R}^2$ (or in $\mathbb{R}^3$) with a sufficiently smooth boundary S is understood. Sufficiently smooth means that S is C^2–smooth. The S is assumed closed and connected; the origin is assumed to be inside S. By N, the unit normal to S pointing out of D is denoted; by u_N, the normal derivative of u on S is denoted, $D' := \mathbb{R}^3 \setminus D$ is the exterior unbounded region. By $\zeta = \zeta(s)$, the boundary impedance is denoted; $s \in S$ is a point on the boundary.

The problem of wave scattering by an obstacle consists of finding the scattering solution $u(x, \alpha, k)$. The scattering solution solves the problem:

$$\left(\nabla^2 + k^2\right)u = 0 \quad \text{in} \quad D', \tag{1.5}$$

$$u_N + \zeta(s)u = 0 \quad \text{on} \quad S, \tag{1.6}$$

$$u = u_0 + v, \tag{1.7}$$

where u_0 is the plane wave defined in (1.2) and v is the scattered field satisfying the radiation condition (1.3).

The basic assumption on the boundary impedance $\zeta(s)$ is:

$$Im\,\zeta(s) \geq 0. \tag{1.8}$$

The scattering amplitude is defined by formula (1.4). It is known, see, for example, [49], that the obstacle-scattering problem (1.5)–(1.7) has a solution and the solution is unique if condition (1.8) holds.

The problem is to find necessary and sufficient conditions on the scattering amplitude for S to be a sphere and $\zeta(s)$ to be a constant.

In Chapter 3 the following general symmetry problem is studied.

Consider the following over-determined boundary problem for the Helmholtz equation:

$$\left(\nabla^2 + k^2\right)u = c_0 \quad \text{in} \quad D, \tag{1.9}$$

$$u|_S = c_1, \tag{1.10}$$

$$u_N|_S = c_2, \tag{1.11}$$

where c_j, $j = 0, 1, 2$, are constants, $k > 0$ is a fixed constant, D is a bounded domain in $\mathbb{R}^n$ with a C^2-smooth closed connected boundary S, $n = 2$ or $n = 3$.

In general the over-determined problem (1.9)–(1.11) does not have a solution. Assume that it has a solution. Then the question is:

Under what additional assumptions the existence of a solution to problem (1.9)–(1.11) *implies that S is a sphere, that is, that D is a ball?*

In $\mathbb{R}^2$ we call a disc a ball.

Special cases of the above question have been of interest and open for many decades.

For example, if $c_0 = 0$, $c_1 = 0$, and $c_2 = 1$, then one has a problem known as Schiffer's conjecture.

M. Schiffer conjectured that if $c_0 = 0$, $c_1 = 0$, $c_2 = 1$, and problem (1.9)–(1.11) is solvable, then D is a ball.

The author proves this conjecture in Chapter 3 with the following *refinement*:

The radius a of the ball cannot be an arbitrary positive number, it must belong to a countable set, and this set is specified in Chapter 3.

The author proved that $a = a_m$, where a_m belongs to the set of positive solutions to the equation $J_0(ka) = 0$, where $J_0(r)$ is the Bessel function regular at the origin and $n = 2$, and if $n = 3$, then a_m belongs to the set of positive solutions of the equation $j_0(ka) = 0$, where $j_0(r)$ is the spherical Bessel function, $j_0(r) = \sqrt{\frac{\pi}{2r}} J_{1/2}(r)$.

If $c_0 = 0, c_1 = 1$, and $c_2 = 0$, then one has a problem equivalent to the Pompeiu problem; see [27] and references therein, [26], [31], and [33].

One more example, presented in Chapter 3, is an old symmetry problem:

Prove that if $k = 0$, $c_0 = 1$, $c_1 = 0$, and $c_2 = \frac{|V|}{|S|}$, and problem (1.9)–(1.11) is solvable, then D is a ball.

This problem was studied in [56]. It was solved later by a different method in [30]. In Chapter 3 it is studied by a new method.

The general problem (1.9)–(1.11) was not studied in the literature, as far as the author knows.

In Chapter 4 several symmetry problems for partial differential equations (PDEs) are considered.

For example, let

$$u(x) := \int_D \frac{dy}{4\pi|x-y|}, \quad x \in \mathbb{R}^3, \tag{1.12}$$

and assume that

$$u(x) = \frac{c}{|x|}, \quad |x| > R, \tag{1.13}$$

where R is an arbitrary sufficiently large number, such that the ball B_R contains D.

The problem is: derive from these assumptions that D is a ball.

This problem was studied in [1] and [28] by different methods.

In Chapter 4, the above problem is solved by the method developed in [28]. This method allows one to solve other symmetry problems as well; see [26].

A similar problem with the potential (1.12) replaced by a single layer potential

$$v(x) = \int_S \frac{ds}{4\pi|x-s|} \tag{1.14}$$

is considered by a different method; see [6].

Assume that for any harmonic function h the following formula holds:

$$\frac{1}{|S|}\int_S h(s)ds = h(0). \tag{1.15}$$

It is proved in Chapter 4 that then D is a ball.

Another problem studied in Chapter 4 is the following one; see [29]. Suppose one has an obstacle with some boundary condition on F', the large part of the boundary S of the obstacle, and the Dirichlet boundary condition $u|_F = w$ on an arbitrary small part F of the boundary S, namely on $F := S \setminus F'$.

The problem is:

Can one choose w so that the total radiation from the obstacle is less than $\epsilon > 0$, where $\epsilon > 0$ is an arbitrary small number? In other words, can one make an obstacle practically invisible?

This problem is solved in Chapter 4 following paper [29].

In Chapter 5, the Navier–Stokes problem for the whole space $\mathbb{R}^3$ is solved. This is the problem of existence, uniqueness, and continuous dependence on the data of the solution to Navier–Stokes problem in $\mathbb{R}^3$.

This is a nonlinear time-dependent problem, one of the millennium problems, that has been open for many decades; see, for example, [12].

Claude Lois Navier's work that introduced viscosity in the equations of fluid has appeared in 1821 and was developed later by George Gabriel Stokes.

A new *a priori* estimate is the basis for the author's solution of this problem in $\mathbb{R}^3$.

In Chapter 6, the inverse problem of potential theory is solved without the standard assumption about star-shapedness of the bodies. This assumption has been used from 1938, see [15], where it has appeared in the first paper on this subject.

All of the results in this book are taken from the author's papers mentioned in the bibliography.

We do not discuss in this book the author's theory of inverse scattering problems with non-over-determined data (see [44], [45], [46], [49], and [54]), a numerical approach to solving ill-posed problems developed in [47], [48], and the wave scattering by many small impedance particles and applications to creating materials with a desired refraction coefficient (see [37] and [38]).

CHAPTER 2

Necessary and Sufficient Conditions for a Scatterer to be Spherically Symmetric

2.1 SCATTERING BY POTENTIALS

In this chapter the results of the papers [21], [22], and [23] are presented.

Consider the wave-scattering problem by a potential $q(x) \in Q_b$,

$$Q_b := \{q : q = \bar{q}, q \in L^2(B_b), q(x) = 0 \text{ if } |x| > b\}.$$

Let

$$[\nabla^2 + k^2 - q(x)] u = 0 \quad \text{in } \mathbb{R}^3, \tag{2.1}$$

$$u = u_0 + v, \quad u_0 = e^{ik\alpha \cdot x}, \tag{2.2}$$

$$\frac{\partial v}{\partial |x|} - ikv = 0\left(\frac{1}{|x|^2}\right), \quad |x| \to \infty. \tag{2.3}$$

Here $k > 0$ is a wave number, $k = $ const, u_0 is the incident plane wave, $\alpha \in S^2$ is the unit vector in the direction of propagation of this plane wave, S^2 is the unit sphere in $\mathbb{R}^3$. It is well known that problem (2.1)–(2.3) has a solution and this solution is unique; see, for example, [49]. The solution $u = u(x, \alpha, k)$ is called the scattering solution. Condition (2.3) is called the radiation condition. The scattering solution satisfies the following well-known integral equation

$$u(x, \alpha, k) = u_0(x, \alpha, k) - \int_{\mathbb{R}^3} g(x, y) q(y) u(y, \alpha, k) dy, \tag{2.4}$$

where

$$g(x, y) = \frac{e^{ik|x-y|}}{4\pi |x - y|}. \tag{2.5}$$

Since $k > 0$ is fixed, the k–dependence will not be shown in many cases.

It is easy to check by direct differentiation taking into account the equation

$$\left(\nabla^2 + k^2\right) g(x, y) = -\delta(x - y), \tag{2.6}$$

where $\delta(x - y)$ is the delta function, that the solution to (2.4) is the scattering solution, that is, it solves problem (2.1)–(2.3).

The scattered field

$$v(x) = -\int_{\mathbb{R}^3} g(x, y)q(y)u(y, \alpha, k)dy. \tag{2.7}$$

Taking $|x| := r$ to infinity along the direction $\beta := \frac{x}{r}$, one gets

$$v = A(\beta, \alpha, k)\frac{e^{ikr}}{r} + o\left(\frac{1}{r}\right), \tag{2.8}$$

so

$$A(\beta, \alpha, k) = -\frac{1}{4\pi}\int_{B_b} e^{-ik\beta \cdot y}q(y)u(y, \alpha, k)dy. \tag{2.9}$$

The scattering amplitude $A(\beta, \alpha, k)$ is of basic physical interest. It describes the scattering phenomenon. Its properties are studied in detail in the literature, for example, in [49].

The following Lemma, see [49], will be used below.

Lemma 2.1   *If* $q \in C^1(\mathbb{R}^3) \cap Q_b$, *then*

$$u(x, \alpha, k) = u_0(x, \alpha, k)[1 + o(1)] \quad as \quad |x| \to \infty. \tag{2.10}$$

Here $o(1)$ *is unform with respect to* $\alpha \in S^2$.

It is clear that for any vector $\xi \in \mathbb{R}^3$ one can find $\alpha \in S^2$ and $\beta \in S^2$ such that

$$\lim_{k \to \infty} [k(\alpha - \beta)] = \xi. \tag{2.11}$$

Here one can change α and β as $k \to \infty$ and keep these vectors so that

$$\frac{\alpha - \beta}{|\alpha - \beta|} = \frac{\xi}{|\xi|}, \quad |\alpha - \beta| = k^{-1}.$$

Let us define the Fourier transform by the formula

$$\tilde{f} = \int e^{-ik\xi \cdot x} f(x)dx, \quad \int := \int_{\mathbb{R}^3}. \tag{2.12}$$

Lemma 2.2   *If* $\tilde{f}(\xi) = \tilde{f}(|\xi|)$, *then* $f(x) = f(|x|)$.

Proof of Lemma 2.2. For simplicity let us assume that $f \in C^1(\mathbb{R}^3) \cap Q_b$. Denote $|\xi| := \rho$, $|x| := r$, $x^0 := \frac{x}{r}$. By the inversion formula for the Fourier transform, assuming that $\tilde{f}(\xi) = \tilde{f}(|\xi|)$ one has:

$$
\begin{aligned}
f(x) &= \frac{1}{(2\pi)^3} \int f(|\xi|)e^{i\xi \cdot x} d\xi = \frac{1}{(2\pi)^3} \int_0^\infty d\rho \rho^2 \tilde{f}(\rho) \int_{S^2} d\xi^0 e^{i\rho\xi^0 \cdot x^0 r} \\
&= \frac{1}{2\pi} \int_0^\infty d\rho \rho^2 \tilde{f}(\rho) j_0(r\rho) = f(r) = f(|x|).
\end{aligned}
\tag{2.13}
$$

Here the following known formula was used; see, for example, [49]:

$$
\int_{S^2} e^{ir\rho\xi_0 \cdot x_0} d\xi_0 = (4\pi)^2 j_0(r\rho).
\tag{2.14}
$$

Lemma 2.2 is proved. □

Let us now give a proof of the following result.

Theorem 2.3 *Assume that $q \in C^1(\mathbb{R}^3) \cap Q_b$. For q to be spherically symmetric, it is necessary and sufficient that*

$$
A(\beta, \alpha, k_0) = A(\beta \cdot \alpha, k_0), \quad \forall \alpha, \beta \in S^2,
\tag{2.15}
$$

where $k_0 > 0$ is is a fixed number.

Proof of Theorem 2.3. First, according to the author's uniqueness theorem, proved in [49], the knowledge of $A(\beta, \alpha, k_0)$ for all $\alpha, \beta \in S^2$ determines uniquely $q(x) \in C^1(\mathbb{R}^3) \cap Q_b$, and, therefore, determines uniquely $A(\beta, \alpha, k)$ for all $k > 0$.

Substitute formula (2.8) into formula (2.7), choose α and β so that $k(\alpha - \beta)$ is directed along a vector ξ, use formula (2.9) and let $k \to \infty$. From (2.9) and (2.10) one obtains the following formula:

$$
A(\beta, \alpha, k) = -\frac{1}{4\pi} \int_{\mathbb{R}^3} e^{ik(\alpha - \beta) \cdot y} q(y) dy + o(1), \quad k \to \infty,
$$

and taking $k \to \infty$, yields the desired result:

$$
\lim_{k \to \infty; k(\alpha - \beta) = \xi} A(\beta, \alpha, k) = -\frac{1}{4\pi} \tilde{q}(\xi).
\tag{2.16}
$$

Note that

$$
k^2 |\alpha - \beta|^2 = k^2 (2 - 2\alpha \cdot \beta) = |\xi|^2.
\tag{2.17}
$$

Therefore

$$
\alpha \cdot \beta = 1 - \frac{|\xi|^2}{2k^2},
\tag{2.18}
$$

so

$$A(\beta,\alpha,k) = A\left(1 - \frac{|\xi|^2}{2k^2}, k\right). \tag{2.19}$$

It follows from (2.16) and (2.19) that

$$\tilde{q}(\xi) = \tilde{q}(|\xi|). \tag{2.20}$$

By Lemma 2.2, formula (2.20) implies

$$q(x) = q(r), \quad r = |x|. \tag{2.21}$$

The sufficiency part of Theorem 2.3 is proved. □

Let us prove the *necessity* part.

If condition (2.21) holds, then by separation of variables in Equation (2.1) in the spherical coordinates one gets

$$u(x,\alpha,k) = \sum_{\ell=0}^{\infty} R_\ell(kr) P_\ell(\alpha \cdot \beta), \quad \beta = \frac{x}{r} = x^0, \tag{2.22}$$

where R_ℓ are the radial parts of the scattering solution, $P_\ell(\alpha \cdot \beta)$ are the Legendre polynomials; see, for example, [9]. Therefore formula (2.15) holds and the necessity part of Theorem 2.3 is proved. □

Theorem 2.3 is proved. □

Let us formulate the fundamental uniqueness theorem from [49] that we have used.

Theorem 2.4 *If $q \in Q_b$, then the values of the scattering amplitude $A(\beta,\alpha,k_0)$ at a fixed $k = k_0 > 0$ determines uniquely $A(\beta,\alpha,k)$ for all $k > 0$.*

Remark 2.5 It is also proved in [49] that if $q \in Q_b$, then the values of $A(\beta,\alpha_0,k)$ for all $\beta \in S^2$, a fixed $\alpha = \alpha_0 \in S^2$ and all $k > 0$ determine $q(x)$ uniquely.

Actually a stronger result is proved in [49]:

The values of the scattering amplitude known for β belonging to an open subset of S^2, fixed $\alpha = \alpha_0$ and k belonging to an open subset of $(0, \infty)$ determine $Q \in Q_b$ uniquely.

Let us now formulate a general problem. Consider the group $SO(3)$ of rotations R in $\mathbb{R}^3$ and let $Rq := q(R^{-1}x)$.

The problems we are interested in can be stated as follows:

(a) *What are the transformations of the scattering amplitude $A(\beta, \alpha, k)$ induced by the transformation $R : q \rightarrow Rq$?*

(b) *What are the necessary and sufficient conditions on the scattering amplitude for the scatterer to be spherically symmetric?*

By a scatterer we mean a potential $q \in Q_b$ or an obstacle, that is, a bounded connected domain D with a C^2–smooth boundary S.

Denote the scattering amplitude corresponding to a potential $q \in Q_b$ by A_q. Consider a rotation of the coordinate system x and let $y = Rx$, so that $x = R^{-1}y$. Every vector α in the coordinate system x after the rotation becomes $R\alpha$ in the coordinate system y. The scattering amplitude $A_q(\beta, \alpha, k)$ in the coordinate system y becomes $A_{Rq}(R\beta, R\alpha, k)$.

Equation (2.1) becomes

$$\left[\nabla^2 + k^2 - q(R^{-1}y)\right]u = 0, \tag{2.23}$$

because the operator ∇^2 is invariant with respect to rotations.

The incident plane wave $e^{ik\alpha \cdot x}$ is invariant under rotations because $\alpha \cdot x$ is. Indeed,

$$\alpha \cdot x = R\alpha \cdot Rx = R^{-1}R\alpha \cdot x = \alpha \cdot x, \tag{2.24}$$

because $R' = R^{-1}$, where $R' = R^*$, R^* is the adjoint matrix of the rotation R and R' is the transposed matrix of this rotation. The factor $\frac{e^{ik|x|}}{|x|}$ is also invariant under rotations because $|x|$ is. Thus, we have proved the following result.

Theorem 2.6   *One has:*

$$A_q(\beta, \alpha, k) = A_{Rq}(R\beta, R\alpha, k). \tag{2.25}$$

Since $R\beta \cdot R\alpha = \beta \cdot \alpha$ one concludes that the condition $A(\beta, \alpha, k) = A(\beta \cdot \alpha, k)$ and formula (2.25) imply

$$A_q(\beta \cdot \alpha, k) = A_{Rq}(R\beta \cdot R\alpha, k) = A_{Rq}(\beta \cdot \alpha, k). \tag{2.26}$$

If $q \in Q_b$, then $Rq \in Q_b$. By Theorem 2.4 it follows that if $q \in Q_b$ and (2.26) holds for $k = k_0 > 0$, then it holds for all $k > 0$ and

$$q(x) = Rq(x), \tag{2.27}$$

that is

$$q(x) = q\left(R^{-1}x\right) \quad \forall R \in SO(3). \tag{2.28}$$

This means that q is spherically symmetric.

Using (2.28) we prove Theorem 2.7.

Theorem 2.7   *Assume $q \in Q_b$. Then q is spherically symmetric if and only if*

$$A_q(\beta, \alpha, k_0) = A_q(\beta \cdot \alpha, k_0), \quad k_0 > 0. \tag{2.29}$$

Here k_0 is a fixed number.

***Proof of Theorem* 2.7.** It follows from Theorem 2.4 that if $q \in Q_b$, then (2.29) implies

$$A_q(\beta, \alpha, k) = A_q(\beta \cdot \alpha, k) \quad \forall k > 0. \tag{2.30}$$

If q is spherically symmetric, then we have already proved by a reference to separation of variables in the spherical coordinates that condition (2.29) holds.

This proves the necessity of the condition (2.29) *for the spherical symmetry of the potential.*

If condition (2.29) holds, then condition (2.30) holds.

If one assumes that $q \in Q_b$ has additional smoothness, for example, $q \in C^1(\mathbb{R}^3)$, then one can use Theorem 2.3 and conclude that q is spherically symmetric.

If one only assumes that $q \in Q_b$, then one gets

$$A_{Rq}(R\beta, R\alpha, k_0) = A_{Rq}(R\beta \cdot R\alpha, k_0) = A_{q(R^{-1}x)}(\beta \cdot \alpha, k_0). \tag{2.31}$$

By Theorem 2.6, this implies

$$A_{q(R^{-1}x)}(\beta \cdot \alpha, k_0) = A_q(\beta \cdot \alpha, k_0). \tag{2.32}$$

By Theorem 2.4, relation (2.32) implies

$$q(x) = q(R^{-1}x) \quad \forall R \in SO(3). \tag{2.33}$$

This means that q is spherically symmetric.

The sufficiency of the condition (2.29) *for the spherical symmetry of the potential is proved.*

Thus, Theorem 2.7 is proved. $\qquad\qquad\square$

Remark 2.8   In the proof of Theorem 2.7, no smoothness of q is assumed and condition (2.29) holds for a single $k = k_0$.

2.2 SCATTERING BY OBSTACLES

Consider now wave scattering by an obstacle D. We assume that D is a bounded connected domain with C^1−smooth boundary S. By $D' := \mathbb{R}^3 \setminus D$ the exterior unbounded region is denoted, N is the unit outer normal to S and $\zeta(s)$ is the boundary impedance. We assume that

$$Im\zeta(s) \geq 0. \tag{2.34}$$

The obstacle-scattering problem consists of finding the scattering solution, that is, the solution to the equation:

$$(\nabla^2 + k^2)u = 0 \quad \text{in} \quad D', \tag{2.35}$$

$$u_N + \zeta(s)u = 0 \quad \text{on} \quad S, \tag{2.36}$$

$$u = u_0 + v, \tag{2.37}$$

where $u_0 = e^{ik\alpha \cdot x}$ is the incident plane wave and v is the scattered field satisfying the radiation condition (2.3) at infinity.

It is well known that the scattering problem (2.35)–(2.37) has a solution and this solution is unique if condition (2.34) holds; see, for example, [49]. Condition (2.34) is important for the uniqueness of the scattering solution; see Lemma 2.11 below.

The scattering amplitude $A_{S,\zeta} := A_{S,\zeta}(\beta, \alpha, k)$ is defined by formula (2.8). We prove the following result.

Theorem 2.9   *If and only if S is a sphere and $\zeta(s) = const$ the scattering amplitude has the property*

$$A_{S,\zeta}(\beta, \alpha, k) := A_{S,\zeta}(\beta \cdot \alpha, k) \quad \forall k > 0. \tag{2.38}$$

The proof of Theorem 2.9 requires a formula similar to (2.26) and the uniqueness theorem, Theorem 2.10 below. Theorem 2.10 is proved in [49].

Theorem 2.10   *If*

$$A_{S_1, \zeta_1(s)}(\beta, \alpha, k_0) = A_{S_2, \zeta_2(s)}(\beta, \alpha, k_0) \quad \forall \alpha, \beta \in S^2, \tag{2.39}$$

then

$$S_1 = S_2, \quad \zeta_1(s) = \zeta_2(s). \tag{2.40}$$

Proof of Theorem 2.9. If S is a sphere and $\zeta(s) = const$, Im $\zeta \geq 0$, then relation (2.38) holds as one checks by separation of variables in the spherical coordinates. This proves the *necessity* of the condition (2.38) for the spherical symmetry of the scatterer, that is, for the S to be a sphere and $\zeta(s)$ to be a constant.

To prove the *sufficiency*, assume that condition (2.38) holds. Then the argument used in the proof of Theorem 2.7 yields the conclusion that S is a sphere and $\zeta(s) = const$. This proves the sufficiency of the condition (2.30). In this argument, Theorem 2.10 was used and the following formula

$$A_{S,\zeta(s)}(\beta, \alpha, k_0) = A_{RS, R\zeta(s)}(R\beta, R\alpha, k_0), \tag{2.41}$$

where $R\zeta(s) = \zeta(R^{-1}s)$. $\qquad\qquad\square$

Let us prove for the convenience of the reader the uniqueness result (Theorem 2.10), mentioned above.

Let $B_R; = \{x : |x| \leq R\}$, $B_R' := \mathbb{R}^3 \setminus B_R$, and assume that $D \subset B_R$, $D_R := D \cap B_R$. By $\bar{v}$ the complex conjugate of v is denoted.

Lemma 2.11 *If v solves the scattering problem with the incident field u_0 equal to zero and Im $\zeta(s) \geq 0$, then $v = 0$.*

Proof of Lemma 2.11. Consider the identity

$$
\begin{aligned}
0 &= \int_{D_R'} \left[\bar{v} \left(\nabla^2 + k^2 \right) v - v \left(\nabla^2 + k^2 \right) \bar{v} \right] dx \\
&= \int_{S_R} (\bar{v} v_N - v \bar{v}_N) \, ds - \int_S (\bar{v} v_N - v \bar{v}_N) \, ds \\
&= 2ik \int_{S_R} |v|^2 ds + 2i \int_S \text{Im } \zeta(s) |v|^2 ds.
\end{aligned}
\tag{2.42}
$$

Here we have used the radiation condition for v and the impedance boundary condition.

If $\text{Im} \zeta(s) \geq 0$, then (2.42) implies

$$
\int_{S_R} |v|^2 ds = 0, \quad \text{Im } \zeta(s) = 0.
\tag{2.43}
$$

From the first equation (2.43), it follows that $v = 0$ in B_R'. By the unique continuation property for solutions to the homogeneous Helmholtz equation, it follows that $v = 0$ in D'.

Lemma 2.11 is proved. □

Let D_1 and D_2 be two obstacles, $D_1 \cap D_2 = D^{12}$, $D_1 \cup D_2 = D_{12}$, $\partial D^{12} = S^{12}$, $\partial D_{12} = S_{12}$. Let G be a connected component of $D_{12} \setminus D^{12}$, $G' := D_1 \setminus G$, $G' \subset D_1$, $S' = \partial G$ be the boundary of G, $S_1' = S' \cap S_1$, $S_2' = S' \cap S_2$, $A_j(\beta, \alpha, k)$, $j = 1, 2$, be the scattering amplitudes corresponding to scattering problems for D_1 and, respectively, for D_2.

Lemma 2.12 *If*
$$
A_1(\beta, \alpha, k_0) = A_2(\beta, \alpha, k_0) \quad \forall \alpha, \beta \in S^2,
\tag{2.44}
$$
then $S_1 = S_2$ and $\zeta_1(s) = \zeta_2(s)$.

Proof of Lemma 2.12. If (2.44) holds, then $u_1 - u_2 = o(|x|^{-1})$ as $|x| \to \infty$ and u_j, $j = 1, 2$, are the scattering solutions. By Lemma 1.2.1 from [49], p. 30, it follows that $u_1 = u_2$ in $D_{12}' = \mathbb{R}^3 \setminus D_{12}$. Therefore $u_1 = u_2$ on S_1' and $u_{1N} = u_{2N}$ on S_1'. By the unique continuation property for solutions to homogeneous Helmholtz equation, the function u_1 admits the unique continuation into G as the solution to the Helmholtz equation. This continuation equals to u_2 in D_2'. Let us denote by w this continuation. Then

$$
(\nabla^2 + k^2) w = 0 \quad \text{in} \quad G,
\tag{2.45}
$$

$$(w_N + \zeta_1(s)w)\,|_{S'_1} = 0, \quad (-w_N + \zeta_2(s)w)\,|_{S'_2} = 0. \tag{2.46}$$

Here $-w_N$ appears because on S'_2 the normal N is directed into G.

Consider the identity:

$$\begin{aligned}
0 &= \int_G \left[\bar{w}\left(\nabla^2 + k^2\right)w - w\left(\nabla^2 + k^2\right)\bar{w}\right] ds = \int_{S'}\left(\bar{w}w_N - \bar{w}_N w\right) ds \\
&= \int_{S'_1}\left(-\zeta_1(s)|w|^2 + \bar{\zeta}_1(s)|w|^2\right) ds + \int_{S'_2}\left(\bar{\zeta}_2(s)|w|^2 - \zeta_2(s)|w|^2\right) ds.
\end{aligned} \tag{2.47}$$

If we assume that $\operatorname{Im}\zeta_j > 0$, $j = 1, 2$, on a set of positive surface measure on S', then it follows from (2.47) that $w = 0$ on S'_1. This and the impedance boundary conditions for u_j allow one to conclude that $u_{1N} = 0$ on S'_1. By the uniqueness of the solution to the Cauchy problem for the Helmholtz equation, the relation $u_1 = u_{1N} = 0$ on S'_1 imply $u_1 = 0$ in G. Therefore $u_1 = 0$ in D'_1. This is impossible because $|u_1(x, \alpha, k_0)| \to 1$ as $|x| \to \infty$. This contradiction proves that $D_1 = D_2$ and $S_1 = S_2$. It also proves that $\zeta_1(s) = \zeta_2(s)$ because $\zeta_j(s) = \frac{u_{jN}(s)}{u_j(s)}$, $j = 1, 2,$, and $u_1 = u_2$ in D', where $D := D_1 = D_2$.

Therefore, Lemma 2.12 is proved under the assumption that $\operatorname{Im}\zeta_j(s) > 0$ on a set of positive surface measure for at least one of the j.

If $\operatorname{Im}\zeta_j(s) = 0$, then the above argument proves the existence of the non-trivial solutions to the boundary problem for $w = w(x, \alpha, k_0)$ with $k_0 := k$:

$$\left(\nabla^2 + k^2\right)w = 0 \quad \text{in} \quad G, \tag{2.48}$$

$$(w_N + \zeta_1(s)w)\,|_{S'_1} = 0, \quad (-w_N + \zeta_2(s)w)\,|_{S'_2} = 0. \tag{2.49}$$

Choose distinct $\alpha_m \in S^2$, $m = 1, 2, \ldots, M$, so that functions $w(x, \alpha_m, k_0)$ are linearly independent in G. This is possible to do with M as large as we wish because this is possible to do for $u_0(x, \alpha_m, k_0)$. Then one has M eigenfunctions of the Laplacian $\mathcal{L}$ defined in G by the impedance boundary conditions. Since the resolvent of this Laplacian is compact (because G is a bounded domain), the operator $\mathcal{L}$ has discrete spectrum and it has finitely many eigenvalues not exceeding k^2. This is a contradiction since M can be chosen as large as one wishes. This contradiction proves Lemma 2.12. $\qquad\square$

Exercise 2.1 Prove that the set w_m, $1 \le m \le M$, is linearly independent in $L^2(G)$ if $\alpha_p \ne \alpha_q$ for $p \ne q$.

Hint: see [49, pp. 99–100].

CHAPTER 3

Symmetry Problems for the Helmholtz Equation

The contents of this chapter is based on the author's papers [55] and [51].

3.1 A GENERAL SYMMETRY PROBLEM

In this chapter we study the over-determined boundary problem for the Helmholtz equation:

$$(\nabla^2 + k^2) u = c_0 \quad \text{in} \quad D \tag{3.1}$$

$$u|_S = c_1, \tag{3.2}$$

$$u_N|_S = c_2. \tag{3.3}$$

Here c_j, $j = 0, 1, 2$, are constants, $k > 0$ is a constant, D is a bounded connected C^2–smooth domain with boundary S. We assume that u and the constants c_j are real-valued. This is the case in the applications discussed in this chapter. Since $k^2 > 0$, this assumption can be done without loss of generality.

In general, problem (3.1)–(3.3) does not have a solution.

Suppose that a solution to the above problem does exist. Then the problem is:

Find necessary and sufficient conditions for D to be a ball.

By a ball in the two-dimensional space $\mathbb{R}^2$ we understand a disc.

Let $J_0(r)$ be the Bessel function regular at the origin, and $j_0(r)$ be the spherical Bessel function, $j_0(r) = \left(\frac{\pi}{2r}\right)^{1/2} J_{1/2}(r)$.

Our basic result is the following theorem.

Theorem 3.1 *If problem* (3.1)–(3.3) *has a solution and*

$$\left| c_1 - \frac{c_0}{k^2} \right| + |c_2| > 0, \tag{3.4}$$

then D is a ball B_a of radius a, where a solves the equation

$$J_0'(ka) = \frac{C_2}{kC_1} J_0(ka), \tag{3.5}$$

if $D \subset \mathbb{R}^2$.

Here

$$C_1 := c_1 - \frac{c_0}{k^2}, \quad C_2 := c_2. \tag{3.6}$$

If $D \subset \mathbb{R}^3$, then a solves the equation

$$j_0'(ka) = \frac{C_2}{kC_1} j_0(ka). \tag{3.7}$$

Let us first reduce our problem to the problem with $c_0 = 0$.
Define

$$v = u + c_3, \quad c_3 := -\frac{c_0}{k^2}. \tag{3.8}$$

Then our problem can be written as

$$\left(\nabla^2 + k^2\right) v = 0, \tag{3.9}$$

$$v|_S = C_1, \quad v_N|_S = C_2. \tag{3.10}$$

Thus, problem (3.1)–(3.3) is reduced to an equivalent problem (3.9)–(3.10) with the homogeneous Helmholtz equation.

Remark 3.2   If $|C_1| + |C_2| = 0$, then problem (3.9)–(3.10) has a solution $v = 0$ for any shape of D. This makes the condition (3.4) necessary and sufficient for the existence of the solution to problem (3.1)–(3.3) to imply that D is a ball.

Below we discuss mainly problem (3.9)–(3.10).

Its special case, corresponding to $C_1 = 0$, $C_2 = 1$, has been an open problem for many decades. It is known as Schiffer's conjecture.

We prove the *refined Schiffer's conjecture*. The refinement consists of the specification of the set of the radii of the ball: These radii are not arbitrary but belong to a specific set. The result is formulated as Theorem 3.3 below.

Theorem 3.3   *Assume that*

$$|C_1| + |C_2| > 0 \tag{3.11}$$

and problem (3.9)–(3.10) is solvable. Then D is a ball B_a with radius $a = a_m$, $m = 1, 2, \ldots$, where a_m are positive roots of the equation

$$J_0(ka) = 0, \quad D \subset \mathbb{R}^2. \tag{3.12}$$

If $D \subset \mathbb{R}^3$, then a_m are positive roots of the equation

$$j_0(ka) = 0. \tag{3.13}$$

If $C_1 = 1$, $C_2 = 0$, then problem (3.9)–(3.10) is equivalent to the modern formulation of *the Pompeiu problem*; see [26] and references therein, [31], [33], and [36]. Our result is the following theorem.

Theorem 3.4   *If $C_1 = 1$, $C_2 = 0$ and problem* (3.9)–(3.10) *is solvable, then D is a ball B_{a_m} where the admissible radii a_m are positive roots of the equation*

$$J_0'(ka) = 0, \quad D \subset \mathbb{R}^2. \tag{3.14}$$

If $D \subset \mathbb{R}^3$, then a_m are positive roots of the equation

$$j_0'(ka) = 0. \tag{3.15}$$

Theorem 3.5   *If $C_1 = 0$, $C_2 = 1$ and problem* (3.9)–(3.10) *is solvable, then D is a ball B_{a_m} where the admissible radii a_m are positive roots of the equation*

$$J_0(ka) = 0 \quad D \subset \mathbb{R}^2. \tag{3.16}$$

If $D \subset \mathbb{R}^3$, then a_m are positive roots of the equation

$$j_0(ka) = 0. \tag{3.17}$$

Theorem 3.5 contains a solution to the refined Schiffer's conjecture.

Let us prove these theorems.

Proof of Theorem 3.1. Let us first prove that if problem (3.9)–(3.10) is solvable and condition (3.11) holds, then D is a ball. We assume that $D \subset \mathbb{R}^2$.

Recall some known facts. Let $r = r(s)$ be the parametric equation of S and s be the arclength of S. We assume that the origin is inside D. Then

$$\frac{dr}{ds} = t(s), \tag{3.18}$$

where t is the unit vector tangent to S at the point s. We say point s meaning point $\mathbf{r}(s)$. Furthermore,

$$\frac{dt}{ds} = k(s)v(s), \tag{3.19}$$

where $k(s) \geq 0$ is the curvature of S at the point s and $v(s)$ is the unit normal to S at the point s. If S is convex, then v is directed inside D. The normal v changes direction (sign) when s

passes through the points at which the convexity of S changes sign. If S is strictly convex, then $k(s) > 0 \; \forall s \in S$.

If $u_N = C_2 \neq 0$ and N is the unit outer normal to S, then the sign of convexity of S is unchanged, and if $C_2 > 0$, then

$$v(s) = -N(s), \quad k(s) > 0 \quad \text{if} \quad C_2 > 0. \tag{3.20}$$

The boundary condition (3.10) implies

$$u(x(s), y(s)) = C_1, \quad r(s) = x(s)e_1 + y(s)e_2, \tag{3.21}$$

where $\{e_1, e_2\}$ is a Cartesian orthonormal basis of $\mathbb{R}^2$.

Differentiate (3.21) with respect to S and get

$$v_x x' + v_y y' = 0, \tag{3.22}$$

where

$$x' := \frac{dx}{ds} = t_1(s), \quad y' := \frac{dy}{ds} = t_2(s). \tag{3.23}$$

Differentiate (3.22) and get

$$v_{xx} t_1^2 + 2v_{xy} t_1 t_2 + v_{yy} t_2^2 + v_x x_{ss} + v_y y_{ss} = 0. \tag{3.24}$$

Using formulas (3.18)–(3.19) one obtains

$$v_{xx} t_1^2 + 2v_{xy} t_1 t_2 + v_{yy} t_2^2 = k(s)C_2, \quad C_2 > 0. \tag{3.25}$$

Here the following relations were used:

$$v_x x_{ss} + v_y y_{ss} = \nabla v \cdot \frac{dt}{ds} = -k(s)C_2, \tag{3.26}$$

and we assumed $C_2 > 0$. Consequently $N = -v$.

It follows from (3.25) that v_{xx}, v_{xy} and v_{yy} cannot vanish simultaneously on S since $C_2 \neq 0$ and $k(s) > 0$. The differential Equation (3.9) implies that

$$v_{xx} + v_{yy} = -k^2 C_1. \tag{3.27}$$

Therefore, if $C_1 \neq 0$, then v_{xx} and v_{yy} cannot vanish simultaneously on S.

Denote

$$v_{xx}|_S = p(s), \quad v_{xy}|_S = q(s). \tag{3.28}$$

Then

$$v_{yy}(s) = -k^2 C_1 - p(s). \tag{3.29}$$

Let A be a 2×2 symmetric matrix with the elements

$$A_{11} = p(s), \quad A_{12} = A_{21} = q(s), \quad A_{22} = -k^2 C_1 - p(s). \tag{3.30}$$

Equation (3.25) can be written as

$$(At, t) = k(s)C_2, \tag{3.31}$$

where

$$t = t_1 e_1 + t_2 e_2, \quad t \cdot z := (t, z) = t_1 z_1 + t_2 z_2, \tag{3.32}$$

and $t \cdot z$ is the scalar product of two vectors. One has

$$\|t\|^2 = (t, t) = t_1^2 + t_2^2. \tag{3.33}$$

Let us calculate eigenvalues of A. They are roots of the equation

$$\det(A - \lambda I) = 0. \tag{3.34}$$

Here I is the identity matrix.

By (3.30), Equation (3.34) can be written as

$$\lambda^2 + k^2 C_1 \lambda - p^2(s) - q^2(s) - k^2 C_1 p(s) = 0. \tag{3.35}$$

Therefore, the eigenvalues are:

$$\lambda_{1,2} = -\frac{k^2 C_1}{2} \pm \left(\left(\frac{k^2 C_1}{2} \right)^2 + p^2(s) + q^2(s) + k^2 C_1 p(s) \right)^{1/2}, \tag{3.36}$$

or

$$\lambda_{1,2} = -\frac{k^2 C_1}{2} \pm \left(\left(\frac{k^2 C_1}{2} + p(s) \right)^2 + q^2(s) \right)^{1/2}. \tag{3.37}$$

Let us calculate the eigenvectors W_1 and W_2 corresponding to these eigenvalues.

The eigenvector W_1, corresponding to the eigenvalue λ_1, is found from the equation

$$\det(A - \lambda_1 I)W_1 = 0.$$

Let $W_{11}, W_{1,2}$ be the coordinates of W_1 in the basis $\{e_1, e_2\}$. Then

$$\begin{aligned} (p(s) - \lambda_1)W_{11} + q(s)W_{12} &= 0 \\ q(s)W_{11} + (-k^2 C_1 - p(s) - \lambda_1)W_{12} &= 0. \end{aligned} \tag{3.38}$$

The eigenvector is determined up to a constant factor. Therefore one may choose $W_{11} = 1$. With this choice one has $W_{12} = \frac{\lambda_1 - p(s)}{q(s)}$ as follows from the first Equation (3.38) if $q(s) \neq 0$.

If $q(s) = 0$, then one may take $W_{12} = 0$.

Thus, one obtains

$$W_1 = \{1, \gamma\}, \quad \gamma = \frac{\lambda_1 - p(s)}{q(s)}, \quad q(s) \neq 0, \tag{3.39}$$

$$W_1 = \{1, 0\}, \quad \text{if} \quad q(s) = 0. \tag{3.40}$$

Similarly one finds W_2:

$$W_2 = \{-\gamma, 1\}, \tag{3.41}$$

where

$$W_{21} = \frac{k^2 C_1 + p(s) + \lambda_2}{q(s)},$$

and one checks that

$$-\frac{k^2 C_1 + p(s) + \lambda_2}{q(s)} = \frac{\lambda_1 - p(s)}{q(s)}, \quad q(s) \neq 0. \tag{3.42}$$

To verify (3.42) it is sufficient to check that

$$-k^2 C_1 = \lambda_1 + \lambda_2. \tag{3.43}$$

Relation (3.43) follows immediately from (3.35).

If $q(s) = 0$, then

$$W_1 = \{1, 0\}, \quad W_2 = \{0, 1\}. \tag{3.44}$$

From (3.39)–(3.41) it follows that

$$(W_1, W_2) = 0, \tag{3.45}$$

and

$$\|W_j\|^2 = 1 + \gamma^2, \quad j = 1, 2. \tag{3.46}$$

Since the length of a vector is invariant with respect to rotations of the coordinate system, the quantity γ^2 is also invariant with respect to such rotations.

Let us express vectors $t(s)$ in terms of the eigenvectors W_1 and W_2. This is possible since vectors W_1 and W_2 are orthogonal, non-zero, and, therefore, form a basis of $\mathbb{R}^2$. One has

$$t = h_1 W_1 + h_2 W_2, \tag{3.47}$$

where h_j are some constant coefficients. To find h_j one writes (3.47) as a linear algebraic system

$$\begin{aligned} W_{11} h_1 + W_{21} h_2 &= t_1, \\ W_{12} h_1 + W_{22} h_2 &= t_2. \end{aligned} \tag{3.48}$$

The determinant Δ of this system is

$$\Delta := W_{11} W_{22} - W_{12} W_{21} = 1 + \gamma^2, \tag{3.49}$$

where formulas (3.39) and (3.41) were used.

Since $\Delta > 0$, the linear algebraic system (3.48) is uniquely solvable. This system can be written as

$$h_1 - \gamma t_2 = t_1, \quad \gamma h_1 + h_2 = t_2. \tag{3.50}$$

One can solve this system explicitly:

$$h_1 = \frac{t_1 + \gamma t_2}{\Delta}, \quad h_2 = \frac{t_2 - \gamma t_1}{\Delta}. \tag{3.51}$$

Substitute (3.47) into (3.31) and take into account that

$$AW_j = \lambda_j W_j, \quad (W_i, W_j) = \delta_{ij} \Delta. \tag{3.52}$$

The result is:

$$(h_1 AW_1 + h_2 AW_2, \, h_1 W_1 + h_2 W_2) = h_1^2 \lambda_1 \Delta + h_2^2 = k(s)C_2. \tag{3.53}$$

Using formulas (3.51) rewrite (3.53) as

$$\lambda_1 (t_1 + \gamma t_2)^2 + \lambda_2 (t_2 - \gamma t_1)^2 = k(s)C_2\Delta. \tag{3.54}$$

Let us prove that (3.54) may hold only if D is a ball. It follows from (3.37) that

$$\lambda_2 < 0 \quad \text{if} \quad C_1 > 0, \tag{3.55}$$

and

$$\lambda_1 > 0 \quad \text{if} \quad C_1 < 0. \tag{3.56}$$

Note that one can rotate the coordinate system so that the ratio $\frac{t_1}{t_2}$ may take any desired value from $(-\infty, \infty)$. The only restriction on $\{t_1, t_2\}$ is $t_1^2 + t_2^2 = 1$.

Choose the coordinate system so that $t_1 = 0$, $t_2 = 1$. Then (3.54) yields

$$\lambda_1 \gamma^2 + \lambda_2 = k(s)C_2\Delta. \tag{3.57}$$

Now choose the coordinate system so that $t_1 = 1$, $t_2 = 0$. Then (3.54) yields

$$\lambda_1 + \gamma^2 \lambda_2 = k(s)C_2\Delta. \tag{3.58}$$

The right side in (3.57)–(3.58) is invariant with respect to rotations.

Therefore from (3.57)–(3.58) one obtains

$$(\lambda_1 - \lambda_2)\left(1 - \gamma^2\right) = 0. \tag{3.59}$$

Consequently,

$$\lambda_1 = \lambda_2 := \lambda = -\frac{k^2 C_1}{2}, \tag{3.60}$$

or

$$\gamma^2 = 1. \tag{3.61}$$

If (3.60) holds, then one derives from (3.37) that

$$q(s) = 0, \quad \frac{k^2 C_1}{2} + p(s) = 0, \quad \gamma = 0, \quad \Delta = 1. \tag{3.62}$$

From (3.62) and (3.54) it follows that

$$\lambda \left(t_1^2 + t_2^2\right) = \lambda = -\frac{k^2 C_1}{2} = k(s)C_2. \tag{3.63}$$

Therefore

$$k(s) = -\frac{k^2 C_1}{2C_2}. \tag{3.64}$$

The right side of (3.64) does not depend on s, so the curvature of S is constant; it does not depend on s.

Thus, S is a sphere in $\mathbb{R}^2$.

Assume now that (3.61) holds. The case $\gamma = -1$ is treated in the same way as the case $\gamma = 1$. Therefore, assume that $\gamma = 1$. Then it follows from (3.54) that

$$\lambda_1 \left(t_1 + t_2\right)^2 + \lambda_2 \left(t_2 - t_1\right)^2 = 2k(s)C_2. \tag{3.65}$$

This implies:

$$\lambda_1 + \lambda_2 + 2t_1 t_2 \left(\lambda_1 - \lambda_2\right) = 2k(s)C_2. \tag{3.66}$$

Take in (3.66) $t_1 = 0$ and use the relation $\lambda_1 + \lambda_2 = -k^2 C_1$ to get (3.64).

Therefore, the curvature $k(s)$ is a constant.

This implies that S is a sphere in $\mathbb{R}^2$.

If S is a sphere, then D is a ball B_a of radius a. In this case Equation (3.9) has a solution $c J_0(kr)$, $c = const$, $r = |x|$. The boundary conditions (3.10) are satisfied if $c J_0(ka) = C_1$ and $ck J_0'(ka) = C_2$. Consequently,

$$J_0'(ka) = \frac{C_2}{kC_1} J_0(ka). \tag{3.67}$$

It is known, see, for example, [11], that this equation has countably many positive roots a_m. The set of these roots is the set of the admissible radii in Theorem 3.3. Theorems 3.1 and 3.3 are proved. $\qquad\square$

Proof of Theorems 3.4 and 3.5. The conclusion of Theorem 3.4 follows from Theorem 3.3 with $C_1 = 1$, $C_2 = 0$.

The conclusion of Theorem 3.5 follows from Theorem 3.3 with $C_1 = 0$, $C_2 = 1$.

Theorems 3.4 and 3.5 are proved. $\qquad\square$

Exercise 3.1 Prove Theorems 3.1–3.5 for the case $D \subset \mathbb{R}^3$.

Hint. First prove that a normal section of S is a circle.

Then prove that if any normal section of S for a fixed normal is a circle, then S is a sphere. See also Theorem 3.8 in Section 3.3.

3.2   OLD SYMMETRY PROBLEM

Consider now an old symmetry problem. It can be considered as problem (3.1)–(3.3) with $k = 0$, $c_0 = 1, c_1 = 0\, c_2 = |D|/|S|$, where $|D|$ is the volume of D and $|S|$ is the area of S. The known result is formulated in the following theorem.

Theorem 3.6   *If problem* (3.1)–(3.3) *with $k = 0$, $c_0 = 1$, $c_1 = 0\, c_2 = |D|/|S|$ is solvable, then D is a ball.*

This result was proved in [56], by a different method in [30], and by another method in [51]. Our new proof is based on the same ideas as the proof of Theorem 3.1.

Proof of Theorem 3.6. We give a proof assuming that $D \subset \mathbb{R}^2$.

Let $r(s) = x(s)e_1 + y(s)e_2$ be parametric representation of S, $r(s)$ is the radius vector corresponding to parameter s, and s is the arc length of S, the natural parameter. Since

$$u(x(s, y(s)) = 0,$$

one can differentiate this condition with respect to s and get

$$u_x x' + u_y y' = 0.$$

Differentiate this relation again and get

$$u_{xx} t_1^2 + 2u_{xy} t_1 t_2 + u_{yy} t_2^2 = fk(s), \quad f := |D|/|S|. \tag{3.68}$$

Let

$$u_{xx}(s) = p(s) := p, \quad u_{xy}(s) = q(s) := q. \tag{3.69}$$

Denote by A the 2×2 symmetric matrix with the elements

$$A_{11} = p, \quad A_{12} = A_{21} = q, \quad A_{22} = 1 - p. \tag{3.70}$$

Here we have used differential Equation (3.1) on S and took into account that $k = 0$.

Let $t = \{t_1, t_2\}$. Write (3.68) as

$$(At, t) = fk(s). \tag{3.71}$$

Let us find eigenvalues and eigenvectors of A. The eigenvalues are roots of the characteristic equation $\det(A - \lambda I) = 0$. Using (3.70), one writes the characteristic equation $\det(A - \lambda I) = 0$ as

$$\lambda^2 - \lambda - p^2 - q^2 + p = 0. \tag{3.72}$$

Thus, the eigenvalues are:

$$\lambda_{1,2} = \frac{1}{2} \pm \left[\left(\frac{1}{2} - p \right)^2 + q^2 \right]^{1/2}. \tag{3.73}$$

One has

$$\lambda_1 + \lambda_2 = 1, \quad \lambda_1 \lambda_2 = -p^2 - q^2 + p. \tag{3.74}$$

The corresponding eigenvectors are:

$$W_1 = \{1, \gamma\}, \quad \gamma = \frac{\lambda_1 - p}{q}, \quad q \neq 0, \tag{3.75}$$

$$W_2 = \{-\gamma, 1\}. \tag{3.76}$$

If $q = 0$, then one may take $\gamma = 0$.

Using relations similar to (3.47)–(3.54), one derives the relation

$$\lambda_1 (t_1 + \gamma t_2)^2 + \lambda_2 (t_2 - \gamma t_1)^2 = fk(s)\Delta, \quad \Delta = 1 + \gamma^2. \tag{3.77}$$

It follows from (3.73) that $\lambda_1 > 0$. Choose a coordinate system in which $t_1 = 0, t_2 = 1$. Then formula (3.77) yields

$$\lambda_1 \gamma^2 + \lambda_2 = fk(s)\Delta. \tag{3.78}$$

In the coordinate system in which $t_1 = 1, t_2 = 0$, relation (3.77) yields

$$\lambda_1 + \lambda_2 \gamma^2 = fk(s)\Delta. \tag{3.79}$$

The quantities γ^2, f, and the curvature $k(s)$ are invariant with respect to rotations of the coordinate system, s being fixed.

From (3.78)–(3.79) one obtains:

$$(\lambda_1 - \lambda_2) \Delta = 0. \tag{3.80}$$

Thus, $\lambda_1 = \lambda_2 = \frac{1}{2}$. Therefore, by formula (3.73) it follows that $q^2 + (\frac{1}{2} - p)^2 = 0$, so

$$q = 0, \quad p = \frac{1}{2}, \quad \gamma = 0, \quad \Delta = 1. \tag{3.81}$$

If $\gamma = 0$ and $\lambda_1 = \lambda_2 = \frac{1}{2}$, formula (3.77) yields

$$\frac{1}{2} = fk(s), \tag{3.82}$$

so

$$k(s) = \frac{1}{2f}. \tag{3.83}$$

Since $f = const$, it follows that $k(s) = const$. This implies that S is a sphere in $\mathbb{R}^2$ and D is a ball B_a.

In this case, $u = \frac{|x|^2 - a^2}{4}$ and $u_N|_S = \frac{a}{2} = \frac{|D|}{|S|}$.

Theorem 3.6 is proved. $\square$

3.3   NECESSARY AND SUFFICIENT CONDITIONS FOR S TO BE A SPHERE

Let us give some necessary and sufficient conditions for a smooth connected surface S in $\mathbb{R}^3$ to be a sphere. By $r = r(p,q)$, we denote a parametric equation of S, by N, the unit outer normal to S is denoted, $[r, N]$ is the vector product of two vectors, and $r_p = \frac{\partial r}{\partial p}$.

Theorem 3.7   *If and only if*

$$[r, N]|_S = 0 \tag{3.84}$$

the surface S is a sphere.

Proof of Theorem 3.7. *The necessity* of the condition (3.84) is obvious since the radius-vector of a point on a sphere is directed along the normal to the sphere at this point.

To prove *the sufficiency* of the condition (3.84) for S to be a sphere, recall that $[r_p, r_q]$ is a vector directed along the normal N to S.

Condition (3.84) can be written as

$$[r, [r_p, r_q]] = 0, \tag{3.85}$$

or

$$r_p(r, r_p) - r_q(r, r_p) = 0. \tag{3.86}$$

Vectors r_p and r_q are linearly independent if S is a smooth surface. Therefore (3.86) implies

$$(r, r_p) = 0, \quad (r, r_q) = 0. \tag{3.87}$$

Consequently,

$$(r, r) = const. \tag{3.88}$$

This means that S is a sphere. Theorem 3.7 is proved.   □

Let P be a normal section of D, that is, the intersection of D with the plane containing a normal $N = N_s$ to S at some point s. We say normal sections for a fixed N, meaning normal sections containing this N only.

Theorem 3.8   *If all the normal sections of D for a fixed N are discs, then S is a sphere. Conversely, if S is a sphere, then all its normal sections for a fixed N are discs.*

Proof of Theorem 3.8. Choose a point $s \in S$ and assume that all normal sections P_s for the normal N_s are discs. Take one of these discs. Let O be the center of this disc, a be its radius, and AB be its diameter. The other normal section for this normal is also a disc with the same center, the same diameter, and the same radius. Therefore $[r, N_s] = 0$ for all $s \in S$. By Theorem 3.7, one concludes that S is a sphere. Theorem 3.8 is proved.   □

3.4   THE POMPEIU PROBLEM

Let us briefly discuss the Pompeiu problem. Dimitrie Pompeiu (1873–1954) was a Romanian mathematician, a student of H. Poincare. He published in 1929 a paper (see [16]) in which the following question was discussed. Suppose that

$$\int_{\sigma(D)} f(x)dx = 0 \quad \forall \sigma \in G, \tag{3.89}$$

where G is the group of rigid motions in $\mathbb{R}^2$, that is, the group of rotations and translations.

It is assumed that D is a bounded connected domain with a boundary S and $f \in L_{loc}^1(\mathbb{R}^2 \cap \mathbb{S}')$, where $\mathbb{S}'$ is the Schwartz space of distributions. In [16] it was assumed that $f \in C(\mathbb{R}^2)$ locally.

The problem was:

Does (3.89) *imply that* $f = 0$?

D. Pompeiu in [16] claimed that (3.89) implies that $f = 0$. This was wrong. A counterexample was found in [3] 15 years later: If D is a disc, then f can be not zero and (3.89) still holds.

The modern problem is:

Is disc the only exceptional domain? In other words, if (3.89) *holds and D is not a disc, does it follow that $f = 0$?*

This problem has been open for many decades; see [31]. It was proved in [33] that the answer is yes. In this section we give a new proof of this result. Our proof is based on Theorem 3.1.

We reduce this problem to Theorem 3.4. To do this, let us start with rewriting (3.89) as

$$\int_D f(Rx + y)dx = 0 \quad \forall y \in \mathbb{R}^2 \quad \forall R \in SO(2). \tag{3.90}$$

If (3.90) implies $f = 0$, we say that D has P−property. Otherwise we say that D fails to have P−property.

It will follow from our results that the only domain D which fails to have P−property is a ball.

Theorem 3.9   *If and only if problem*

$$\left(\nabla^2 + k^2\right)u = -1 \quad in \quad D \subset \mathbb{R}^2, \tag{3.91}$$

$$u\,|_S = u_N|_S = 0, \tag{3.92}$$

where $k > 0$ is a constant, has a solution, the domain D fails to have $P-$ property.

Proof of Theorem 3.9. If u solves the over-determined problem (3.91)–(3.92), then $v := u + \frac{1}{k^2}$ solves the problem

$$\left(\nabla^2 + k^2\right)v = 0, \quad v|_S = \frac{1}{k^2}, \quad v_N|_S = 0. \tag{3.93}$$

This is the problem equivalent to the one discussed in Theorem 3.4. From Theorem 3.4, it follows that if problem (3.93) is solvable, then D is a ball of radius a_m, where a_m solves the equation $J_0'(ka) = 0$.

Conversely, if D is a ball of this radius, then a separation of variables shows that problem (3.93) is solvable.

Theorem 3.9 is proved. □

Let us give additional information, see also [26].

Assume that D fails to have P−property. Then there is an $f \neq 0$ such that (3.90) holds. Let

$$\tilde{f}(\xi) = \int f(x)e^{ix\cdot\xi}dx, \quad \int := \int_{\mathbb{R}^n}, \quad n \geq 2. \tag{3.94}$$

Taking the Fourier transform of (3.90), one gets

$$\tilde{f}(\xi)\overline{\tilde{\chi}(R^{-1}\xi)} = 0, \quad \forall R \in SO(n), \tag{3.95}$$

where the overline denotes complex conjugate, $\chi(x)$ is the characteristic function of D, that is, $\chi(x) = 1,\ x \in D$ and $\chi(x) = 0\ x \notin D$.

Let $\mathcal{N} := \cap \mathcal{N}_R$, where the intersection is taken over all rotations R and $\mathcal{N}_R = \{\xi : \tilde{\chi}(R^{-1}\xi) = 0\}$. By definition, $\mathcal{N}$ is rotationally invariant. Since D is bounded, $\tilde{\chi}(\xi)$ is an entire function of ξ. It follows from (3.95) that $\tilde{f}(\xi)$ has support $\mathcal{N}$. If D fails to have P−property, that is, $f \neq 0$, then $\tilde{f} \neq 0$, so $\mathcal{N}$ is not an empty set. Since $\mathcal{N}$ is rotationally invariant, it contains a circle S_a of radius $a > 0$. Since $\tilde{\chi}(\xi) = 0$ on S_a, one has

$$\tilde{\chi}(\xi) = (\xi^2 - a^2)\,\tilde{u}(\xi). \tag{3.96}$$

Taking the inverse Fourier transform of (3.96), one gets Equation (3.91) with $k^2 = a^2$. Since $\tilde{\chi}(\xi)$ is an entire function vanishing on the algebraic variety $\xi^2 - a^2 = 0$, it follows that $\tilde{u}(\xi)$ is an entire function. Therefore $u(x)$ is compactly supported by the Paley–Wiener theorem. So, $u(x) = 0$ for $|x| > b$, where $b > 0$ is sufficiently large. Since $u \in H_{loc}^2(\mathbb{R}^n)$, it follows that $u = u_N = 0$ on S. These are conditions (3.92). We have proved that if D fails to have P−property, then problem (3.91)–(3.92) has a solution.

Conversely, if this problem has a solution, then define $w = 0$ in D and $w = 0$ in $D' = \mathbb{R}^n \setminus D$. This w solves the equation $(\nabla^2 + k^2)w = -\chi$ in $\mathbb{R}^n$. Taking the Fourier transform, one gets $(-\xi^2 + k^2)\tilde{w} = -\tilde{\chi}$. Thus, $\tilde{\chi} = 0$ on the algebraic variety $\xi^2 = k^2$. Consequently, for all rotations $\tilde{\chi}(\xi)$ is invariant, so $\chi(x)$ is also invariant. Thus, (3.95) holds and (3.92) is verified with $f \neq 0$.

CHAPTER 4

Other Symmetry Problems

The contents of this chapter is based on the author's papers [28], [29], and [6].

4.1 VOLUME POTENTIAL

Consider the following problem. Denote

$$u(x) = \int_D g_0(x, y)dy, \quad g_0(x, y) := \frac{1}{4\pi|x - y|}. \tag{4.1}$$

Here $D \subset \mathbb{R}^3$ is a bounded connected domain with C^2—smooth boundary S.

Assume that

$$u(x) = \frac{c}{|x|}, \quad |x| \geq R, \quad c = const, \tag{4.2}$$

where $R > 0$ is an arbitrary large number such that $D \subset B_R$, B_R is a ball of radius R centered at the origin.

Problem 1: Does (4.2) imply that D is a ball?

Theorem 4.1 *If* (4.2) *holds, then D is a ball.*

Proof of Theorem 4.1. By direct calculation, one checks that

$$u_a(x) := \int_{B_a} \frac{1}{4\pi|x - y|}dy = \frac{c_a}{|x|}, \quad |x| > a, \quad c_a = \frac{a^3}{3}. \tag{4.3}$$

It is easy to see that

$$u(x) = \frac{|D|}{4\pi|x|} + O\left(\frac{1}{|x|^2}\right), \tag{4.4}$$

where $|D|$ is the volume of D. From (4.3) and (4.4) it follows that

$$c = \frac{|D|}{4\pi}. \tag{4.5}$$

If

$$a^3 = \frac{3|D|}{4\pi}, \tag{4.6}$$

then the potential $u(x)$ is equal to the potential of the ball B_a for $|x| > R$.

One has

$$\nabla^2 u = -\chi(x), \tag{4.7}$$

where $\chi(x) := \chi_D(x)$ is the characteristic function of D, and

$$\nabla^2 u_a = -\chi_{B_a}(x). \tag{4.8}$$

Since

$$u(x) = u_a(x) \text{ for } |x| > R, \tag{4.9}$$

it follows by the unique continuation property for harmonic functions that

$$u(x) = u_a(x) \text{ for } x \in \mathbb{R}^3 \setminus (D \cup B_a). \tag{4.10}$$

Let $\mathcal{H}$ denote the set of all harmonic functions in B_R. Multiply (4.7) by a harmonic in B_R function $h \in \mathcal{H}$ and integrate over B_R to get

$$\int_{B_R} h(x)\nabla^2 u(x)dx = -\int_D h(x)dx \quad \forall h \in \mathcal{H}. \tag{4.11}$$

Integrating by parts, one obtains

$$\int_{B_R} h(x)\nabla^2 u \, dx = \int_{S_R} (hu_r - h_r u) \, ds, \tag{4.12}$$

where $u_r := \frac{\partial u}{\partial r}$, $r := |x|$. Since $u = \frac{c}{r}$ for $r \geq R$, it follows from (4.12) that

$$\int_D h \, dx = \frac{c}{R} \int_{S_R} h_r \, ds + \frac{c}{R^2} \int_{S_R} h \, ds. \tag{4.13}$$

Since h is harmonic in B_R, one has

$$\int_{S_R} h_r \, ds = 0, \quad \frac{1}{4\pi R^2} \int_{S_R} h \, ds = h(0). \tag{4.14}$$

Thus,

$$\int_D h(x)dx = 4\pi c h(0) \quad \forall h \in \mathcal{H}. \tag{4.15}$$

Denote by $\mathcal{R}$ a rotation, an element of $SO(3)$ the group of rotations in $\mathbb{R}^3$. It is known and can be easily verified that if $h(x) \in \mathcal{H}$, then $h(\mathcal{R}x) \in \mathcal{H}$.

If $\mathcal{R}(\phi)$ is the rotation about a unit vector ℓ by the angle ϕ, then

$$\frac{d\mathcal{R}(\phi)x}{d\phi}\Big|_{\phi=0} = [\ell, x], \tag{4.16}$$

where $[\ell, x]$ is the vector product of two vectors. A derivation of this formula can be found, for example, in [28, p. 51]. Substitute in (4.15) in place of h the function $h(\mathcal{R}(\phi)x)$, differentiate with respect to ϕ, and then set $\phi = 0$. The result is:

$$\int_D \nabla h \cdot [\ell, x] dx = 0 \quad \forall \ell \in S^2 \quad \forall h \in \mathcal{H}, \tag{4.17}$$

where S^2 is the unit sphere in $\mathbb{R}^3$.

Note that

$$\nabla h \cdot [\ell, x] = \nabla \cdot (h[\ell, x]). \tag{4.18}$$

Applying the divergence theorem to (4.17), one gets

$$\int_S h(s) N \cdot [\ell, s] ds = 0 \quad \forall \ell \in S^2, \quad \forall h \in \mathcal{H}. \tag{4.19}$$

Therefore,

$$\int_S h(s)[s, N] ds = 0 \quad \forall h \in \mathcal{H}. \tag{4.20}$$

We now use the following lemma.

Lemma 4.2 *The set of restrictions of all harmonic functions on S is dense in $L^2(S)$.*

Proof of Lemma 4.2 is given after the proof of Theorem 4.1.

By Lemma 4.2, it follows from (4.20) that

$$[s, N] = 0 \text{ on } S. \tag{4.21}$$

From (4.21) and Theorem 3.7 in Chapter 3 it follows that S is a sphere, so D is a ball. Theorem 4.1 is proved. $\qquad\square$

Proof of Lemma 4.2. Assuming that the set of restrictions harmonic functions is not dense in $L^2(S)$, one finds $f \neq 0$, $f \in L^2(S)$, such that

$$\int_S f(s)h(s) ds = 0 \quad \forall h \in \mathcal{H}. \tag{4.22}$$

Choose

$$h(s) = \int_{S_R} \frac{\mu(p) dp}{4\pi |s - p|}, \tag{4.23}$$

where $\mu(p)$ is an arbitrary smooth function on S_R. Then (3.24) yields

$$\int_{S_R} dp \mu(p) \int_S \frac{f(s) ds}{4\pi |s - p|} = 0. \tag{4.24}$$

Since μ is arbitrary, Equation (4.24) implies

$$w(p) := \int_S \frac{f(s)ds}{4\pi|s-p|} = 0 \quad \forall p \in S_R. \tag{4.25}$$

Since w is a single layer potential vanishing on S_R and at infinity, it vanishes everywhere in the region $|x| \geq R$ and, therefore, $w = 0$ in $D' := \mathbb{R}^3 \setminus D$. So $w|_S = 0$, $\nabla^2 w = 0$ in D. Consequently,

$$w = 0 \text{ in } D \cup D', \tag{4.26}$$

and by the jump formula for the normal derivative of a single layer potential (see, for example, [49, p. 18]), one obtains

$$f = w_N^+ - w_N^- = 0, \tag{4.27}$$

where $w_N^\pm$ are the limiting values of the normal derivatives on S from inside and outside of D. Thus, $f = 0$ on S. Lemma 4.2 is proved.   □

4.2   SURFACE POTENTIAL

Let us consider a similar problem in which the volume potential is replaced by a surface potential.
   Assume that the origin is inside D and the following condition holds:

$$v(x) = \int_S \frac{ds}{4\pi|x-s|} = \frac{c}{|x|}, \quad |x| \geq R, \ c = const. \tag{4.28}$$

The problem is:
   Does (4.28) *imply that S is a sphere?*
   The answer is given in the following theorem.

Theorem 4.3   *If* (4.28) *holds, then S is a sphere.*

Prof of Theorem 4.3. One has

$$v(x) = \frac{|S|}{4\pi|x|} + O\left(|x|^{-2}\right), \quad |x| \to \infty. \tag{4.29}$$

Therefore the constant c in (4.28) is

$$c = \frac{|S|}{4\pi}. \tag{4.30}$$

By the unique continuation property for harmonic functions if (4.28) holds in the region $|x| \geq R$, then it holds in $D' = \mathbb{R}^3 \setminus D$, so $v(x) = \frac{|S|}{4\pi|x|}$ in D'.
   By the jump relation formula for the normal derivative of the single layer potential, one has

$$v_N^+ - v_N^- = 1. \tag{4.31}$$

From the assumption (4.28), it follows that

$$v_{\bar{N}} = -\frac{c\,N_s \cdot s}{|s|^3}.$$ (4.32)

If S is not a sphere, then there is a point $s_0 \in S$ such that

$$|s_0| \leq |s| \quad \forall s \in S.$$ (4.33)

At this point N_s is directed along the radius-vector s_0, so (4.32) yields

$$v_{\bar{N}_{s_0}} = -\frac{|S|}{4\pi|s_0|^2}.$$ (4.34)

By the isoperimetric inequality one has

$$|S| > 4\pi|s_0|^2,$$ (4.35)

so

$$v_{\bar{N}_{s_0}} < -1.$$ (4.36)

From (4.31) and (4.36) it follows that

$$v_{\overset{+}{N}_{s_0}} < 0.$$ (4.37)

One has

$$v(s)|_S = \frac{|S|}{4\pi|s|} \leq \frac{|S|}{4\pi|s_0|}.$$ (4.38)

Therefore the harmonic function $v(x)$ is continuous in D and attains its maximum at the point $s_0 \in S$. Consequently,

$$v_{\overset{+}{N}_{s_0}} \geq 0,$$ (4.39)

which contradicts (4.37). This contradiction proves that S is a sphere.

Theorem 4.3 is proved. $\qquad\square$

From Theorem 4.3 one can derive the following result.

Theorem 4.4 *Assume that*

$$\frac{1}{|S|}\int_S h(s)\,ds = h(0) \quad \forall h \in \mathcal{H}.$$ (4.40)

Then S is a sphere.

Proof of Theorem 4.4. Take in (4.40) $h(s) = \frac{1}{|x-s|}$, $x \in B'_R := \mathbb{R}^3 \setminus B_R$. This h is harmonic in D. Substitute it into (4.40) and get

$$\frac{1}{|S|} \int_S \frac{1}{|x-s|} ds = \frac{1}{|x|}, \quad x \in B'_R. \tag{4.41}$$

Consequently,

$$\int_S \frac{ds}{4\pi|x-s|} = \frac{|S|}{4\pi|x|} \quad \forall x \in B'_R. \tag{4.42}$$

This relation is the same as (4.28). By Theorem 4.3, one concludes that S is a sphere. Theorem 4.4 is proved. $\qquad\square$

4.3 INVISIBLE OBSTACLES

Consider now the following problem:

Is it possible to have an "invisible" obstacle? In other words, is it possible to have an obstacle such that the total radiation from this obstacle is as small as one wishes at a fixed wave number and a fixed incident wave direction?

This problem has been investigated by the author in [24] and [29].

In order to formulate the problem exactly, let us introduce the necessary definitions.

Let $D \subset \mathbb{R}^3$ be a bounded connected domain with C^2–smooth boundary S, N be a unit outer normal to S, F be a small open subset of S, F' be its complement in S, D' be the complement of D in $\mathbb{R}^3$, S^2 be a unit sphere in $\mathbb{R}^3$, $k > 0$ is fixed and k–dependence is omitted in the scattering amplitude and in the scattering solution, and $\epsilon > 0$ be an arbitrary small fixed number.

Let

$$\left(\nabla^2 + k^2\right) u = 0 \text{ in } D', \tag{4.43}$$

$$u|_F = w, \tag{4.44}$$

$$(u_N + hu)|_{F'} = 0, \tag{4.45}$$

where $h = h(s)$ is a given sufficiently smooth function, Im$h \geq 0$,

$$u = u_0 + v, \quad u_0 = e^{ik\alpha \cdot x}, \quad \alpha \in S^2, \tag{4.46}$$

$$v = A(\beta, \alpha) \frac{e^{ikr}}{r} + O\left(r^{-2}\right), \quad r = |x| \to \infty, \quad \beta = x/r. \tag{4.47}$$

The function $A(\beta, \alpha)$ is the scattering amplitude, its dependence on k is not shown since $k > 0$ is fixed, α is the direction of the incident plane wave u_0, β is the direction of the scattered field.

Define the cross section σ:

$$\sigma = \int_{S^2} |A(\beta)|^2 \, d\beta, \quad A(\beta) := A(\beta, \alpha), \tag{4.48}$$

where α is fixed.

Invisible Obstacle Problem: *Given an arbitrary small $\epsilon > 0$ and an arbitrary small $F \in S$, can one find w so that $\sigma < \epsilon$?*

The answer to this question is *yes*.

Theorem 4.5   *There exists $w \in C_0^\infty(F)$ such that $\sigma < \epsilon$.*

Proof of Theorem 4.5. Let $G(x, y)$ solve the problem:

$$\nabla^2 G(x, y) + k^2 G(x, y) = -\delta(x - y) \text{ in } D', \tag{4.49}$$

$$G\,|_F = 0, \quad (G_N + hG)|_{F'} = 0, \tag{4.50}$$

$$\lim_{|x| \to 0} |x| \left(\frac{\partial G}{\partial |x|} - ikG \right) = 0. \tag{4.51}$$

The author proved in [43] and in [49, p. 52], that

$$G(x, y) = \frac{e^{ikr}}{r} \psi(y, -\beta) + O\left(\frac{1}{r^2}\right), \quad r := |x| \to \infty, \ \beta = x/r. \tag{4.52}$$

Here $\psi(y, \nu)$ is the scattering solution which solves the scattering problem:

$$\left(\nabla^2 + k^2\right) \psi = 0 \text{ in } D', \tag{4.53}$$

$$\psi\,|_F = 0, \quad (\psi_N + h\psi)|_{F'} = 0, \tag{4.54}$$

$$\psi = e^{ik\beta \cdot x} + \eta, \tag{4.55}$$

where η satisfies the radiation condition.

Using G one obtains by Green's formula an equation for v:

$$v(x) = \int_{F'} G(x, s)(u_{0N} + hu_0)ds + \int_F G_N(x, s)v(s)ds. \tag{4.56}$$

It follows from (4.52), (4.56), and (4.52) that

$$A(\beta) = \frac{1}{4\pi} \left[\int_{F'} \psi(s, -\beta)(u_{0N} + hu_0)\, ds + \int_F (w - u_0) \psi_N(s, -\beta)ds \right]. \tag{4.57}$$

One has

$$A = A_0 - A_1, \tag{4.58}$$

where

$$A_0(\beta) = \frac{1}{4\pi} \left[\int_{F'} \psi(s, -\beta)(u_{0N} + hu_0)\, ds - \int_F u_0 \psi_N(s, -\beta)ds \right], \tag{4.59}$$

$$A_1(\beta) = -\int_F w\psi_N(s, -\beta)ds. \tag{4.60}$$

The conclusion of Theorem 4.5 follows immediately from Lemma 4.6. By $\|f\|$ the norm in $L^2(S^2)$ is denoted.

Lemma 4.6 *Given an arbitrary $f \in L^2(S^2)$ and an arbitrary small $\epsilon > 0$, one can find $w \in C_0^\infty(F)$ such that*

$$\|f(\beta) - A_1(\beta)\| < \epsilon. \tag{4.61}$$

We prove this lemma after the proof of Theorem 4.5.

To finish the proof of Theorem 4.5, take as f the function $A_0(\beta)$, use Lemma 4.6 and find $w \in C_0^\infty(F)$ such that

$$\|A_0(\beta) - A_1(\beta)\| < \epsilon. \tag{4.62}$$

Theorem 4.5 is proved. □

Proof of Lemma 4.6. Assume that the conclusion of Lemma 4.6 is false. Then there exists an $f \in L^2(S^2)$ such that

$$\int_{S^2} d\beta f(\beta) \int_F dsw(s)\psi_N(s, -\beta) = 0. \tag{4.63}$$

The set $C_0^\infty(F)$ is dense in $L^2(S^2)$. Therefore, (4.63) implies that

$$\int_{S^2} f(\beta)\psi_N(s, -\beta)d\beta = 0 \quad \forall s \in F. \tag{4.64}$$

Let

$$z(x) := \int_{S^2} f(\beta)\psi(x, -\beta)d\beta, \quad x \in D'. \tag{4.65}$$

Clearly,

$$(\nabla^2 + k^2)z = 0 \text{ in } D', \tag{4.66}$$

$$z|_{x \in F} = 0, \quad z_N|_{x \in F} = 0. \tag{4.67}$$

The conditions (4.67) hold because of (4.54) and (4.64). By the uniqueness of the solution to the Cauchy problem for the elliptic Equation (4.66), it follows that $z(x) = 0$ in D'. Consequently, as we prove below, $f = 0$ and Lemma 4.6 is proved.

Let us derive from the relation $z(x) = 0$ in D' that $f = 0$. One can prove (see [49]) that

$$\psi(x, \beta) = Te^{ik\beta \cdot x}, \tag{4.68}$$

where T is a linear bounded operator acting on x−variable, and T^{-1} is bounded. The specific form of T is not important for our proof. Applying T^{-1} to (4.65) one gets

$$\int_{S^2} f(\beta)e^{-ik\beta \cdot x}d\beta = 0 \quad \forall x \in D'. \tag{4.69}$$

Here the integral is an entire function of x. Therefore

$$\int_{S^2} f(\beta) e^{-ik\beta \cdot x} d\beta = 0 \quad \forall x \in \mathbb{R}^3. \tag{4.70}$$

This means that the Fourier transform of the distribution $f(\beta)\delta(|\xi| - k)|\xi|^{-2}$ vanishes, where $\delta(|\xi| - k)$ is the delta function concentrated on the sphere $|\xi|^2 = k^2$. Thus,

$$f(\beta)\delta(|\xi| - k)|\xi|^{-2} = 0. \tag{4.71}$$

Consequently, $f = 0$.

Lemma 4.6 is proved. $\qquad\qquad\qquad\qquad\qquad\qquad\qquad\qquad\qquad\qquad\qquad\qquad$ $\square$

CHAPTER 5

Solution to the Navier–Stokes Problem

In this chapter, the author's papers [34] and [35] are used.

We prove that the Navier–Stokes (NS) problem in $\mathbb{R}^3$ has a smooth solution and this solution is unique if the data are smooth and rapidly decaying.

This is a millennium problem [12]. There is a very large literature on this subject. We mentioned only the books [7], [12], and references therein.

The results of this chapter are obtained in [34], [35], and our presentation follows very closely these papers. These results are described in [53].

Let us state the Navier–Stokes equations in $\mathbb{R}^3$; see [8]. Let $v = v(x, t)$ be the velocity of the fluid, $x = (x_1, x_2, x_3) \in \mathbb{R}^3$. The NS equation is:

$$v_t + (v, \nabla)v = -\nabla p + \nu \Delta v + f, \quad \text{in} \quad \mathbb{R}^3, \quad \nabla \cdot v = 0,$$

where p is the pressure, f is the force, ν is the (kinematic) viscosity, and the density of the fluid is a constant which we take equal to 1 without loss of generality.

As noted, there is a large literature on the Navier–Stokes (NS) problem (see [11], [12], and references therein, including the papers by Leray, Hopf, Lions, Prodi, Kato and others). The global existence and uniqueness of a solution was not proved. Various notions of the solution were used; see [11] and [18].

In this chapter, the solution to NS problem is defined as a solution to an integral equation. It is proved that this solution is unique, it exists in a certain class of functions globally, in other words, for all $t \geq 0$, it has finite energy, and an estimate of this solution is given. This estimate guarantees continuous dependence of the solution on the data in a certain sense. The data are the force f and the initial velocity $v_0(x)$.

The NS problem consists of solving the equations

$$v' + (v, \nabla)v = -\nabla p + \nu \Delta v + f, \quad x \in \mathbb{R}^3, \ t \geq 0, \quad \nabla \cdot v = 0, \quad v(x, 0) = v_0(x). \quad (5.1)$$

Vector-functions $v = v(x, t)$, $f = f(x, t)$ and the scalar function $p = p(x, t)$ decay as $|x| \to \infty$ uniformly with respect to $t \in [0, T]$, where $T > 0$ is an arbitrary large fixed number, $v' = v_t$, $\nu = const > 0$, v_0 is given, $\nabla \cdot v_0 = 0$, the velocity v and the pressure p are unknowns, v_0 and f are known.

Equations (5.1) describe viscous incompressible fluid with density $\rho = 1$. We assume that

$$|f| + |\nabla f| + |v_0| + |\nabla v_0| = O\left(\frac{1}{(1 + |x|)^a}\right) \quad \text{as} \quad |x| \to \infty, \quad a > 3, \qquad (5.2)$$

and for f the decay with respect to x holds for any $t \geq 0$, uniformly with respect to $t \in [0, T]$, where $T \geq 0$ is an arbitrary large fixed number.

5.1 A NEW APPROACH

Our approach consists of three steps.

First, we construct tensor $G := G_{jm}(x, t)$ solving the problem:

$$G' - \nu \Delta G = \delta(x)\delta(t)\delta_{jm} - \nabla p_m, \quad \nabla \cdot G = 0; \quad G = 0 \quad \text{for} \quad t \leq 0. \qquad (5.3)$$

Here $\delta(x)$ is the delta-function, δ_{jm} is the Kronecker delta, $p_m = p(x, t)e_m$, p is the scalar, $p(x, t) = 0, t < 0$; $\nabla p_m = \frac{\partial p}{\partial x_j} e_j e_m$, $\{e_j\}|_{j=1}^3$ is the orthonormal basis of $\mathbb{R}^3$, $e_j e_m$ is a tensor.

Second, we prove that solving problem (5.1) is equivalent to solving the following integral equation:

$$v(x, t) = \int_0^t ds \int_{\mathbb{R}^3} G(x - y, t - s)[f(y, s) - (v, \nabla)v]dy + \int_{\mathbb{R}^3} g(x - y, t)v_0(y)dy, \qquad (5.4)$$

where

$$g(x, t) = \frac{e^{-\frac{|x|^2}{4\nu t}}}{(4\nu t \pi)^{3/2}}, \quad g(x, 0) = \delta(x), \quad g' - \nu \Delta g = \delta(x)\delta(t), \quad \int_{\mathbb{R}^3} g(x, t)dx = 1. \quad (5.5)$$

Third, we prove that Equation (5.4) has a solution in the space X of functions with finite norm $N_1(v) := \sup_{x \in \mathbb{R}^3, t \in [0,T]}\{(|v(x, t)| + |\nabla v(x, t)|)(1 + |x|)\}$ and this solution is unique in X.

The results are stated in Theorems 5.1 and 5.3 in Sections 5.1 and 5.6. The result in Theorem 5.1 is conditional: We assume that $\sup_{n>1} N_1(v_n) < \infty$. The reason is the absence of an *a priori* estimate of the solution v in the space X.

In Theorem 5.7 the crucial *a priori* estimate is derived in the space Y. In Theorem 5.8, the existence, uniqueness, and continuous dependence of the solution on the data are proved in the space Y. The definition of this space is given in Section 5.7. Theorem 5.8 solves one of the millennium problems, the NS problem in $\mathbb{R}^3$.

Let us now state the following theorem in which the existence of the solution to NS problem is conditional.

Theorem 5.1 *Problem (5.1) has a unique solution in X. The solution to (5.1) exists in X provided that* $\sup_{n>1} N_1(v_n) < \infty$, *where v_n is defined in (5.20). This solution has finite energy* $E(t) = \int_{\mathbb{R}^3} |v(x, t)|^2 dy$ *for every $t \geq 0$.*

5.2   CONSTRUCTION OF G

Let

$$G = \int_{\mathbb{R}^3} H(\xi, t) e^{i\xi \cdot x} d\xi.$$

Taking Fourier transform of (5.3) yields

$$H' + \nu\xi^2 H = \frac{\delta(t)\delta_{jm}}{(2\pi)^3} - i\xi P_m(\xi, t), \quad \xi H = 0, \quad H(\xi, t) = 0, \quad t < 0. \tag{5.6}$$

The $H = H_{jm}$ is a tensor, $1 \le j, m \le 3$,

$$P_m(\xi, t) := (2\pi)^{-3} \int_{\mathbb{R}^3} p_m(x, t) e^{-i\xi \cdot x} dx, \quad \xi := \xi_j e_j,$$

summation is understood here and below over the repeated indices, $1 \le j \le 3$, $\xi H := \xi_j H_{jm}$. Multiply (5.6) by ξ from the left, use the equation $\nabla \cdot G = 0$, which implies

$$\xi H = 0, \quad \xi H' = 0,$$

and get

$$(2\pi)^{-3}\delta(t)\xi_m = i\xi^2 P_m, \quad P_m := P_m(\xi, t) = -i\frac{\delta(t)\xi_m}{(2\pi)^3\xi^2}, \quad \xi^2 := \xi_j\xi_j. \tag{5.7}$$

From (5.6)–(5.7) one obtains:

$$H'_{jm} + \nu\xi^2 H_{jm} = (2\pi)^{-3}\delta(t)\left(\delta_{jm} - \frac{\xi_j\xi_m}{\xi^2}\right). \tag{5.8}$$

Thus,

$$H = H_{jm}(\xi, t) = (2\pi)^{-3}\left(\delta_{jm} - \frac{\xi_j\xi_m}{\xi^2}\right)e^{-\nu\xi^2 t}; \quad H = 0 \quad \text{for} \quad t \le 0. \tag{5.9}$$

$$G(x, t) = (2\pi)^{-3}\int_{\mathbb{R}^3} e^{i\xi \cdot x}\left(\delta_{jm} - \frac{\xi_j\xi_m}{\xi^2}\right)e^{-\nu\xi^2 t}d\xi := I_1 + I_2. \tag{5.10}$$

The integrals I_1, I_2 are calculated in Section 5.7:

$$I_1 = \delta_{jm}g(x, t); \quad g(x, t) := \frac{e^{-\frac{|x|^2}{4\nu t}}}{(4\nu t\pi)^{3/2}}, \quad t > 0, \quad g = 0, \ t < 0, \tag{5.11}$$

$$I_2 = \frac{\partial^2}{\partial x_j \partial x_m}\left(\frac{1}{|x|2\pi^{3/2}}\int_0^{|x|/(4\nu t)^{1/2}} e^{-s^2}ds\right), \quad t > 0; \quad I_2 = 0 \quad \text{for} \quad t \le 0. \tag{5.12}$$

Therefore,

$$G(x, t) = \delta_{jm}g(x, t) + \frac{1}{2\pi^{3/2}}\frac{\partial^2}{\partial x_j \partial x_m}\left(\frac{1}{|x|}\int_0^{|x|/(4\nu t)^{1/2}} e^{-s^2}ds\right), \quad t > 0. \tag{5.13}$$

5.3   SOLUTION TO INTEGRAL EQUATION FOR v SATISFIES NS EQUATIONS

Apply the operator $L := \frac{\partial}{\partial t} - v\Delta$ to the left side of (5.4) and use (5.3) to get

$$
Lv = \int_0^t ds \int_{\mathbb{R}^3} \left[\delta(x-y)\delta(t-s)\delta_{jm} + \partial_j \left(\partial_m \left(\frac{1}{2\pi^{3/2}|x-y|} \int_0^{|x-y|/(4v|t-s|)^{1/2}} e^{-p^2} dp \right) \right) \right.
$$
$$
\left. \times (f(y,s) - (v(y,s),\nabla)v(y,s)) \right] dy = f(x,t) - (v(x,t),\nabla)v(x,t) - \nabla p_m(x,t),
$$

$$(5.14)$$

where

$$
p_m(x,t) = -\int_0^t ds \int_{\mathbb{R}^3} \partial_m \left(\frac{1}{2\pi^{3/2}|x-y|} \int_0^{|x-y|/(4v|t-s|)^{1/2}} e^{-p^2} dp \right)
$$
$$
[f(y,s) - (v(y,s),\nabla)v(y,s)]dy
$$

and $v(x,0) = v_0(x)$ because $g(x-y,0) = \delta(x-y)$.

Using the formula $\nabla \cdot G = 0$, the relation

$$
\int_{\mathbb{R}^3} g(x-y,t)v_0(y)dy = \int_{\mathbb{R}^3} g(z,t)v_0(z+x)dz
$$

and the formula $\nabla \cdot v_0 = 0$, one checks that $\nabla \cdot v = 0$.

Thus, a solution to (5.4) *solves* (5.1).

5.4   UNIQUENESS OF THE SOLUTION TO THE INTEGRAL EQUATION

Let X be the space $C^1(\mathbb{R}^3; 1 + |x|)$ of vector-functions with the norm

$$
N_1(v) := \sup_{x\in\mathbb{R}^3, t\in[0,T]} [(|v(x,t)| + |\nabla v(x,t)|)(1 + |x|)],
$$

where $T > 0$ is an arbitrary large fixed number, and let

$$
N_0(v) = \sup_{x\in\mathbb{R}^3} (|v(x,t)| + |\nabla v(x,t)|),
$$

$$
N_0(v) \le N_1(v), \quad |v| := \left(\sum_{j=1}^3 |v_j|^2 \right)^{1/2}.
$$

Assume that there are two solutions v_1 and v_2 to Equation (5.4) with finite norms $N_1(v)$ and let $w := v_1 - v_2$.

One has

$$w(x,t) = \int_0^t ds \int_{\mathbb{R}^3} G(x-y,t-s)\left[(v_2,\nabla)\,v_2 - (v_1,\nabla)\,v_1\right] dy$$

$$= \int_0^t ds \int_{\mathbb{R}^3} G(x-y,t-s)\left[(w,\nabla)v_2 + (v_1,\nabla)\,w\right] dy. \tag{5.15}$$

Therefore

$$N_0(w) \le \int_0^t ds N_0\left(\int_{\mathbb{R}^3} |G(x-y,t-s)|(1+|y|)^{-1}dy\right) N_0(w)\,(N_1(v_2) + N_1(v_1))$$

$$\le c_1 \int_0^t \frac{N_0(w)}{(t-s)^{1/2}} ds, \tag{5.16}$$

where

$$c_1 = \sup_{t\in[0,T]} N_0\left(\int_{\mathbb{R}^3} |G(x-y,t-s)|(1+|y|)^{-1}dy\right)(N_1(v_2) + N_1(v_1)).$$

Since we are proving uniqueness in the set of functions with finite norm $N_1(v)$, one has

$$\sup_{t\in[0,T]} (N_1(v_2) + N_1(v_1)) \le c_2.$$

It is checked in Section 5.7, see formula (5.31), that:

$$N_0\left(\int_{\mathbb{R}^3} |G(x-y,t-s)|(1+|y|)^{-1}dy\right) \le c(t-s)^{-\frac{1}{2}}, \tag{5.17}$$

so inequality (5.16) holds. By c various constants are denoted. From (5.16) by the standard argument, one derives that $N_0(w) = 0$. Thus, $v_1 = v_2$. Uniqueness of the solution to (5.4) is proved in X. $\qquad\square$

5.5 EXISTENCE OF THE SOLUTION TO INTEGRAL EQUATION

Rewrite (5.4) as

$$v(x,t) = -\int_0^t ds \int_{\mathbb{R}^3} G(x-y,t-s)(v,\nabla)v dy + F(x,t), \tag{5.18}$$

where F is known:

$$F(x,t) := \int_0^t ds \int_{\mathbb{R}^3} G(x-y,t-s)f(y,s)dy + \int_{\mathbb{R}^3} g(x-y,t)v_0(y)dy. \tag{5.19}$$

Equation (5.18) is of Volterra type, nonlinear, and uniquely solvable by iterations. The solvability of (5.18) we prove below under assumption (5.21).

One has

$$v_{n+1}(x,t) = -\int_0^t B(v_n, \nabla)v_n ds + F, \quad v_1 = F; \quad Bv := \int_{\mathbb{R}^3} G(x-y, t-s)v(y,s)dy.$$
$$(5.20)$$

Using (5.17), the formula $\int_0^t \frac{ds}{s^{1/2}} = 2t^{1/2}$, and assuming that

$$\sup_{n \geq 1} N_1(v_n) \leq c, \qquad (5.21)$$

one gets:

$$N_0(v_{n+1} - v_n) \leq N_0 \left(\int_0^t ds \int_{\mathbb{R}^3} \frac{|G(x-y,t-s)|}{1+|y|} dy \right) N_0(v_n - v_{n-1})[N_1(v_n) + N_1(v_{n-1})]$$
$$\leq ct^{1/2} N_0(v_n - v_{n-1}). \qquad (5.22)$$

Therefore,

$$N_0(v_{n+1} - v_n) \leq ct^{1/2} N_0(v_n - v_{n-1}). \qquad (5.23)$$

If τ is chosen so that $c\tau^{1/2} < 1$, then B is a contraction map on the bounded set $N_0(v) \leq R$ for sufficiently large R and $t \in [0, \tau]$. Thus, v is uniquely determined by iterations for $x \in \mathbb{R}^3$ and $t \in [0, \tau]$.

The assumption $\sup_{n \geq 1} N_1(v_n) \leq c$ makes the existence result conditional at this moment, but this assumption will be removed later, in Sections 5.8 and 5.9.

If $t > \tau$ rewrite (5.18) as

$$v = F - \int_0^\tau ds \int_{\mathbb{R}^3} G(x-y, t-s)(v, \nabla)v dy - \int_\tau^t ds \int_{\mathbb{R}^3} G(x-y, t-s)(v, \nabla)v dy$$
$$:= F_1 - \int_\tau^t ds \int_{\mathbb{R}^3} G(x-y, t-s)(v, \nabla)v dy,$$

where F_1 is a known function since v is known for $t \in [0, \tau]$. To this equation one applies the contraction mapping principle and get v on the interval $[0, 2\tau]$. Continue in this fashion and construct v for any $t \geq 0$. Process (5.20) converges to a solution to Equation (5.4) under the assumption $\sup_{n \geq 1} N_1(v_n) \leq c$. $\qquad \square$

Remark 5.2   Let us make a remark concerning the mapping done by the operator in (5.18). In [14, p. 234], sufficient conditions are given for a singular integral operator to map a class of functions with a known power rate of decay at infinity into a class of functions with a suitable rate of decay at infinity. It follows from [14], Theorem 5.1 on p. 234, that if

$$|f(x)| + |\nabla f| \leq c(1 + |x|)^{-a}, \quad a > 3,$$

then the part of the operator G in (5.18), responsible for the lesser decay of the iterations v_n at infinity, acts as a weakly singular operator similar to the operator Q,

$$Qf := \int_{\mathbb{R}^3} \frac{b(|x-y|)|f(y)|}{|x-y|^3} dy \le c \frac{1}{(1+|x|)^3}$$

for large $|x|$. This part of G yields the decay $O\left(\frac{1}{(1+|x|)^3}\right)$ at infinity. This decay, in general, cannot be improved. The first iteration v_1 yields a function decaying with its first derivatives as $O\left(\frac{1}{(1+|x|)^3}\right)$. The second iteration contains a function whose behavior for large $|x|$ is determined by the decay of the functions F. Since

$$(v_1 \cdot \nabla) v_1 = O\left((1+|x|)^{-7}\right),$$

the decay of v_2 is again $O\left((1+|x|)^{-3}\right)$ because

$$v_1 = O\left((1+|x|)^{-3}\right) \quad \text{and} \quad |\nabla v_1| = O\left((1+|x|)^{-4}\right).$$

Thus, for the quantity $|v_n| + |\nabla v_n|$, one gets the decay of the order $O\left((1+|x|)^{-3}\right)$ as $|x| \to \infty$ for every fixed $t \ge 0$, provided that f and v_0 together with their first derivatives decay not slower than $O\left((1+|x|)^{-a}\right)$, $a > 3$.

In Sections 5.8–5.9, the existence, uniqueness and continuous dependence of the solution to NS problem in $\mathbb{R}^3$ is proved. The properties of Q allow one to prove that if the smoothness of the data improves, then the smoothness of the solution improves. For infinitely smooth and rapidly decaying data, the solution is infinitely smooth. See also Remark 5.5 below.

5.6   ENERGY OF THE SOLUTION

In this section, we prove that the solution has finite energy in a suitable sense and give an estimate of the solution as $t \to \infty$. This is done under the assumption that $\|\nabla v(x, t)\|^2$ is finite and locally integrable with respect to t. These properties of the solution v are proved below in Section 5.8, where a stronger result is proved, namely, $\sup_{t>0} \|\nabla v(x, t)\|^2 \le c$ provided that the data are sufficiently smooth and decaying fast as $|x| + t \to \infty$.

Let us define the energy by the integral

$$E(t) = \int_{\mathbb{R}^3} |v(x, t)|^2 dy.$$

If one multiplies (5.1) by v and integrate over $\mathbb{R}^3$, then one gets a known conservation law (see [7]):

$$\frac{1}{2} \frac{\partial}{\partial t} \int_{\mathbb{R}^3} |v(x, t)|^2 dx + v \int_{\mathbb{R}^3} \nabla v_j(x, t) \cdot \nabla v_j(x, t) dx = \int_{\mathbb{R}^3} f \cdot v dx, \qquad (5.24)$$

where $f \cdot v$ is the dot product of two vectors.

Integrating (5.24) with respect to t over any finite interval $[0, t]$, 0, and denoting

$$\mathcal{N}(v) := \left(\int_{\mathbb{R}^3} |v(x, t)|^2 dy \right)^{1/2} = \sqrt{E(t)},$$

one gets

$$2\nu \int_0^t \|\nabla v\|^2 dt + E(t) \leq E(0) + 2 \int_0^t ds |\int_{\mathbb{R}^3} f \cdot v dx| \leq E(0) + 2 \int_0^T \mathcal{N}(f)\mathcal{N}(v) ds.$$
$$(5.25)$$

Denote

$$E_T := \sup_{t \in [0,T]} E(t).$$

Maximizing inequality (5.25) with respect to $t \in [0, T]$, and using the elementary inequality $2ab \leq \epsilon a^2 + \frac{1}{\epsilon} b^2$, $a, b, \epsilon > 0$, one derives from (5.25) the following inequality

$$(1 - \epsilon) E_T \leq E(0) + \frac{1}{\epsilon} \left(\int_0^T \mathcal{N}(f) dt \right)^2. \qquad (5.26)$$

Let $\epsilon = \frac{1}{2}$. Then (5.25)–(5.26) allow one to estimate E_T and $\nu \int_0^t \|\nabla v\|^2 dt$ through f and v_0. In particular, we have proved the following theorem (cf [19]).

Theorem 5.3   *Assume that $\int_0^T \mathcal{N}(f) dt < \infty$ for all $T > 0$ and $E(0) < \infty$. Then*

$$\sup_{t \in [0,T]} E(t) < c_{1T} \quad and \quad \int_0^T \|\nabla v(x, t)\|^2 dt < c_{2T}$$

for all $T > 0$. Here $c_{jT} > 0$ are some constants depending on T, $j = 1, 2$.

If $\int_0^\infty \mathcal{N}(f) dt < c$ and $E(0) \leq c$, where $c > 0$ is a constant, then

$$\sup_{t > 0} E(t) < c_1 \quad and \quad \int_0^\infty \|\nabla v(x, t)\|^2 dt < c_2,$$

where $c_j > 0$ are some constants.

Remark 5.4   It is proved in [17] that in a bounded domain $D \subset \mathbb{R}^3$, the solution to a boundary problem for Equations (5.1) in a bounded domain with the Dirichlet boundary condition for large t decays exponentially provided that $\int_0^\infty e^{bt} \mathcal{N}(f) dt < \infty$ for some $b = const > 0$.

5.7   AUXILIARY ESTIMATES

1. *Integral I_1.* One has $I_1 = I_{11}I_{12}I_{13}$, where

$$I_{1p} := \frac{1}{2\pi} \int_{-\infty}^{\infty} e^{ix_p\xi_p - \nu\xi_p^2 t} d\xi_p.$$

No summation over p in the expression $x_p\xi_p$ is understood here.

One has

$$\frac{1}{\pi} \int_0^{\infty} \cos(x_p\xi_p)e^{-\nu\xi_p^2 t} d\xi_p = \frac{1}{\pi}\frac{\pi^{1/2}}{2(\nu t)^{1/2}}e^{-\frac{x_p^2}{4\nu t}} = \frac{e^{-\frac{x_p^2}{4\nu t}}}{(4\nu t\pi)^{1/2}}.$$

After a multiplication (with $p = 1, 2, 3$), this yields formula (5.11) for $g(x, t)$.

Integral I_2. One has

$$I_2 = \frac{\partial^2}{\partial x_j \partial x_m} \frac{1}{(2\pi)^3} \int_{\mathbb{R}^3} e^{i\xi\cdot x} \frac{e^{-\nu t\xi^2}}{\xi^2} d\xi = \frac{\partial^2}{\partial x_j \partial x_m}\left(\frac{1}{(2\pi)^2}\int_0^{\infty} dr e^{-\nu t r^2}\int_{-1}^1 e^{ir|x|u}du\right),$$
$$(5.27)$$

so

$$I_2 = \frac{\partial^2}{\partial x_j \partial x_m}\left(\frac{1}{|x|}\frac{1}{2\pi^{3/2}}\int_0^{\frac{|x|}{(4\nu t)^{1/2}}} e^{-s^2} ds\right). \tag{5.28}$$

Here we have used formula (2.4.21) from [2]:

$$\int_0^{\infty}\frac{\sin(sy)}{s}e^{-as^2}ds = \frac{\pi}{2}Erf(\frac{y}{2a^{1/2}}); \quad Erf(y) := \frac{2}{\pi^{1/2}}\int_0^y e^{-s^2}ds. \tag{5.29}$$

2. One has

$$\int_{\mathbb{R}^3} |\nabla g(x, t)|dx \leq (4\nu t\pi)^{-\frac{3}{2}}\int_{\mathbb{R}^3} e^{-\frac{|x|^2}{4\nu t}}\frac{2|x|}{4\nu t}dx = \frac{2}{(\nu t\pi)^{\frac{1}{2}}}. \tag{5.30}$$

This proves the estimate

$$\int_{\mathbb{R}^3} |G(x - y, t - s)|(1 + |y|)^{-1}dy \leq c(t - s)^{-\frac{1}{2}} \tag{5.31}$$

for the first term of G in (5.13), namely for g. The second term of G is

$$J := \frac{\partial^2}{\partial x_j \partial x_m}\left(\frac{1}{|x|}\int_0^{|x|/(4\nu t)^{1/2}} e^{-s^2} ds\right)$$

up to a factor $\frac{1}{2\pi^{3/2}}$.

One can check by direct differentiation that

$$J = e^{-\frac{|x|^2}{4\nu t}} \left(\frac{\delta_{jm}}{|x|^2 (4\nu t)^{1/2}} - \frac{3x_j x_m}{|x|^4 (4\nu t)^{1/2}} - \frac{2x_j x_m}{|x|^2 (4\nu t)^{3/2}} \right)$$

$$+ \int_0^{|x|/(4\nu t)^{1/2}} e^{-s^2} ds \left(\frac{3x_j x_m}{|x|^5} - \frac{\delta_{jm}}{|x|^3} \right). \tag{5.32}$$

Let,

$$b \left(\frac{|x|}{(4\nu t)^{1/2}} \right) := \int_0^{|x|/(4\nu t)^{1/2}} e^{-s^2} ds.$$

Note that $b(s) = O(s)$ as $s \to 0$ and $b(s) \to \frac{\pi^{1/2}}{2}$ as $s \to \infty$.

It is sufficient to estimate the term with the strongest singularity

$$J_1 := \int_{\mathbb{R}^3} \frac{b(|x - y|)}{|x - y|^3 (1 + |y|)} dy.$$

We prove that

$$\max_{x \in \mathbb{R}^3} J_1 \le c.$$

The other two terms in the second term of J in (5.32) are estimated similarly. Let $y - z = z$. One has

$$J_1 = \int_{\mathbb{R}^3} \frac{b(|z|)}{|z|^3 (1 + |x + z|)} dy = \int_{|z| \le 1} + \int_{|z| \ge 1} := J_{11} + J_{12}.$$

To estimate J_{11} is easy:

$$J_{11} \le c(1 + |x|)^{-1}, \quad x \in \mathbb{R}^3,$$

because

$$b(|z|)|z|^{-3} \le c|z|^{-2} \quad \text{for} \quad |z| \le 1.$$

Let us estimate J_{12}. Note that

$$1 + |y|^a \le (1 + |y|)^a \le 2^{a-1}(1 + |y|^a) \quad \text{for} \quad a \ge 1.$$

We prove that

$$J_{12} \le c \frac{\ln(3 + |x|)}{2 + |x|} \quad \forall x \in \mathbb{R}^3.$$

Use the spherical coordinates $|z| = r$, $|x| := q$. Denote the angle between x and z by θ, $\cos \theta = u$, and get (for $a = 1$) the following estimate:

$$
\begin{aligned}
J_{12} &= 2\pi \int_1^\infty \frac{dr}{r} \int_{-1}^1 \frac{1}{1 + (r^2 + q^2 - 2rqu)^{1/2}} du \\
&\leq c \int_1^\infty \frac{dr}{r} \int_{-1}^1 \frac{1}{(1 + r^2 + q^2 - 2rqu)^{1/2}} du \\
&\leq cq^{-1} \int_1^\infty \frac{dr}{r^2} \frac{(r+q)^2 - (r-q)^2}{(1 + (r+q)^2)^{1/2} + (1 + (r-q)^2)^{1/2}} \\
&\leq c \int_1^\infty \frac{dr}{r(2 + r + q)} \leq c \frac{\ln(3 + q)}{2 + q}.
\end{aligned}
\tag{5.33}
$$

From the above estimates inequality (5.31) follows.

Remark 5.5   In [7] it is shown that the smoothness properties of the solution v are improved when the smoothness properties of f and v_0 are improved.

This also follows from Equation (5.4) by the properties of weakly singular integrals and of the kernel $g(x, t)$.

Let us prove a new *a priori* estimate.

Lemma 5.6   *If the assumptions of Theorem 5.3 hold, then*

$$
(1 + \xi^2) |\tilde{v}(\xi, t)|^2 \leq c \quad \forall t \geq 0,
$$

where $c > 0$ does not depend on ξ and t.

Proof. The Fourier transform of (5.18) yields

$$
\tilde{v}(\xi, t) = \tilde{F}(\xi, t) - \int_0^t ds H(\xi, t - s) \widetilde{(v, \nabla)v}, \quad v(x, t) = \int_{\mathbb{R}^3} e^{i\xi \cdot x} \tilde{v}(\xi, t) d\xi,
\tag{5.34}
$$

where H is defined in (5.9),

$$
|\widetilde{(v, \nabla)v}| = |\tilde{v}_j * i\xi_j \tilde{v}_m| \leq \mathcal{N}(\tilde{v}) \mathcal{N}(|\xi|\tilde{v}),
$$

over the repeated indices j one sums up, $\mathcal{N}(\tilde{v}) := \|\tilde{v}\|_{L^2(\mathbb{R}^3)}$, and $\tilde{v} * \psi$ denotes the convolution of two functions, so

$$
|\widetilde{(v, \nabla)v}| \leq \mathcal{N}(\tilde{v}) \mathcal{N}(|\xi|\tilde{v}) \leq c\mathcal{N}(|\xi|\tilde{v}),
\tag{5.35}
$$

where the known estimate $\mathcal{N}(\tilde{v}) \leq c$ was used. One has

$$
|H| \leq ce^{-\nu t \xi^2}.
\tag{5.36}
$$

It follows from the Parseval's identity that

$$(2\pi)^3 \int_0^t ds \int_{\mathbb{R}^3} |\xi|^2 |\tilde{v}|^2 d\xi = \int_0^t ds \int_{\mathbb{R}^3} |\nabla v|^2 dx \le c, \quad \forall t \ge 0, \qquad (5.37)$$

provided that assumptions of Theorem 5.3 hold.

It follows from (5.34) that

$$|\tilde{v}| \le |\tilde{F}| + c \int_0^t ds e^{-v(t-s)\xi^2} \mathcal{N}(|\xi|\tilde{v})$$

$$\le |\tilde{F}| + c \left(\int_0^t e^{-2v(t-s)\xi^2} ds \right)^{1/2} \left(\int_0^t ds \mathcal{N}^2(\nabla v) \right)^{1/2}. \qquad (5.38)$$

Thus,

$$\left|\xi^2\right| |\tilde{v}|^2 \le 2\left|\xi^2\right| \left|\tilde{F}\right|^2 + 2c^2 \left|\xi^2\right| \frac{1 - e^{-2vt\xi^2}}{2v\xi^2} \le c, \qquad (5.39)$$

provided that

$$\left|\xi^2\right| \left|\tilde{F}\right|^2 \le c.$$

This inequality holds if f and v_0 are smooth and rapidly decaying and if $v \ge c > 0$, which we assume.

We have proved that the inequality $|\tilde{v}| \le ct^{1/2}$ is valid.

Lemma 5.6 is proved. $\qquad \square$

The global existence and uniqueness of a solution to NS problem in $\mathbb{R}^3$ was not yet proved. These properties of the solution are proved in the rest of Chapter 5. One of the new technical points is the usage of the Fourier transform of the unknown velocity and the reduction of the problem to one-dimensional Volterra integral inequalities. The solution to the NS problem is defined as a solution to integral Equation (5.44).

It is proved that this solution is unique, it exists globally, that is, for all $t \ge 0$, and is uniformly bounded in the Sobolev space $H^1(\mathbb{R}^3)$. We assume that the free term f decays sufficiently fast as $|x| + t \to \infty$ and the initial data $v_0(x)$ decays sufficiently fast as $|x| \to \infty$.

Recall that the NS problem in $\mathbb{R}^3$ consists of solving the equations

$$v' + (v, \nabla)v = -\nabla p + v\Delta v + f, \quad x \in \mathbb{R}^3, \ t \ge 0, \quad \nabla \cdot v = 0, \quad v(x,0) = v_0(x). \quad (5.40)$$

Vector-functions $v = v(x,t)$, $f = f(x,t)$ and the scalar function $p = p(x,t)$ are decaying as $|x| \to \infty$ uniformly with respect to $t \in \mathbb{R}_+ := [0, \infty)$, $v' := v_t$, $v = const > 0$ is the viscosity coefficient, the velocity v and the pressure p are unknown, v_0 and f are known.

Equations (5.40) describe viscous incompressible fluid with density $\rho = 1$. By $\bar{v}$, the complex conjugate of v is denoted.

We assume that

$$|f| + |\nabla f| + |v_0| + |\nabla v_0| = O(|x|^{-a}) \quad \text{as} \quad |x| \to \infty,$$

where $a > 3$, and

$$\int_0^\infty \mathcal{N}(f)dt < \infty, \quad \mathcal{N}(f) := \|f\|_{L^2(\mathbb{R}^3)}.$$

We use the integral Equation (5.4)

$$v(x,t) = F - \int_0^t ds \int_{\mathbb{R}^3} G(x-y, t-s)(v, \nabla)v dy. \qquad (5.41)$$

Equation (5.41) is equivalent to (5.40). Formula for the tensor G is derived in Section 5.2. The term $F = F(x,t)$ depends only on the data f and v_0:

$$F := \int_{\mathbb{R}^3} g(x-y)v_0(y)dy + \int_0^t ds \int_{\mathbb{R}^3} G(x-y, t-s)f(y,s)dy. \qquad (5.42)$$

We need an *a priori* estimate for v. We assume throughout that f and v_0 are such that F is bounded in all of the norms we use.

Let Y be the Banach space of continuous functions with respect to t with values in $L^2(\mathbb{R}^3)$ with the norm $\|v\| = \|v\|_X := \sup_{t \in [0,T]} \|v(x,t)\|_{L^2(\mathbb{R}^3)}$, where $T > 0$ is an arbitrary large fixed number.

Denote

$$\mathcal{N}(v) := \|v\|_{L^2(\mathbb{R}^3)}.$$

Then $(v, v) := \mathcal{N}^2(v)$.

In the rest of Chapter 5 we denote by $\mathcal{N}(v)$ the L^2-norm. The reason for choosing L^2-norm is the possibility to derive a new *a priori* estimate in this norm; see Theorem 5.7. In Sections 5.3–5.4, we did not have an *a priori* estimate of the solution in the norm used in these sections. By this reason we had to assume that estimate (5.21) holds. However, with the results of Theorems 5.7–5.8 one could derive *a priori* estimates in the norm used in Sections 5.3–5.4.

The solution to (5.41) is a continuously differentiable function of t with values in the Sobolev space $H^1(\mathbb{R}^3)$ as follows from Theorem 5.7.

Theorem 5.7   *A solution to problem* (5.40) *satisfies the following a priori estimates:*

$$\mathcal{N}(v) \le c_0, \quad \mathcal{N}(\nabla v) \le c(t), \qquad (5.43)$$

where the constant $c_0 > 0$ does not depend on t and $c(t)$ is a continuous function defined for all $t \ge 0$.

The first estimate (5.43) is known, while the second estimate (5.43) is new. This second estimate is crucial for our proofs.

Theorem 5.8   *Problem* (5.40) *has a solution in Y; this solution is unique in Y and depends continuously on the data.*

Remark 5.9   If $\mathcal{N}\left(|\xi|\tilde{F}\right) = O\left(\frac{1}{(t+1)^\gamma}\right)$ as $t \to \infty$ and $\gamma > 1/4$, then $\sup_{t \geq 0} c(t) < \infty$.

This follows from estimate (5.50); see below.

5.8   PROOF OF THE UNIQUENESS OF THE SOLUTION

Proof of the first inequality (5.43).

Multiply (5.40) by $\bar{v}$ and integrate over $\mathbb{R}^3$. Using the standard transformations, one gets

$$2\mathcal{N}(v)\frac{d\mathcal{N}(v)}{dt} + 2v\mathcal{N}^2(\nabla v) = 2Re(f, v).$$

Thus,

$$\frac{d\mathcal{N}(v)}{dt} + v\mathcal{N}^2(\nabla v)/\mathcal{N}(v) \leq \mathcal{N}(f),$$

and

$$\frac{d\mathcal{N}(v)}{dt} \leq \mathcal{N}(f),$$

so

$$\mathcal{N}(v) \leq \int_0^t \mathcal{N}(f)ds + \mathcal{N}(v_0).$$

If

$$\int_0^\infty \mathcal{N}(f)ds < \infty, \quad \mathcal{N}(v_0) < \infty,$$

then the first inequality (5.43) holds.

Proof of the second inequality (5.43).

Let us prove that

$$\psi(t) := \mathcal{N}(\nabla v) \leq c(t).$$

Take the Fourier transform of (5.41), denote $\tilde{v}(\xi, t) := (2\pi)^{-\frac{3}{2}} \int_{\mathbb{R}^3} e^{-i\xi \cdot x} v(x, t)dx$, and let $\tilde{G}$ denote the Fourier transform of G. Then

$$\tilde{v} = \tilde{F} - \int_0^t \tilde{G}(\xi, t - s)\tilde{v} \star (-i\xi\tilde{v})ds := B(\tilde{v}). \tag{5.44}$$

Here $\star$ denotes the convolution in $\mathbb{R}^3$, and for brevity we omitted the tensorial indices: Instead of $\tilde{G}_{mp}\tilde{v}_j \star (-i\xi_j)\tilde{v}_p$, where one sums up over the repeated indices, we wrote $\tilde{G}(\xi, t - s)\tilde{v} \star (-i\xi\tilde{v})$.

Let

$$\psi(t) := \mathcal{N}(\nabla v) = \mathcal{N}(|\xi|\tilde{v}),$$

$$\mathcal{N}(v) = \mathcal{N}(\tilde{v}),$$

where the Parseval's identity is used.

From (5.44) one gets

$$\psi(t) \leq \mathcal{N}\left(|\xi|\tilde{F}\right) + c \int_0^t \mathcal{N}\left(|\xi|\,|\tilde{G}(\xi, t-s)|\right)\psi(s)ds. \tag{5.45}$$

Here and below c stands for various constants independent of t; we have used the *a priori* estimate $\sup_{t\geq 0}\mathcal{N}(\tilde{v}) \leq c_0$ and the standard estimate $|\tilde{v} \star \tilde{w}| \leq \mathcal{N}(\tilde{v})\mathcal{N}(\tilde{w})$. One can check that

$$\mathcal{N}\left(|\tilde{G}(\xi, t-s)|\right) \leq c[v(t-s)]^{-\frac{3}{4}}, \quad \mathcal{N}\left(|\xi||\tilde{G}(\xi, t-s)|\right) \leq c[v(t-s)]^{-\frac{5}{4}}. \tag{5.46}$$

In the derivation of (5.46) we use the estimate $|\tilde{G}(\xi, t)| \leq ce^{-vt\xi^2}$, where $|G|$ is a norm of the matrix; see formula (5.10).

To continue the proof, we need some lemmas.

Lemma 5.10   *The operator $Af := \int_0^t (t-s)^p f(s)ds$ in the Banach space $X_0 := C([0, T])$ has spectral radius $r(A) = 0$ for $p > 0$ and any fixed $T > 0$, $0 \leq t \leq T$.*

Proof. The spectral radius of a bounded linear operator A can be calculated by the formula $r(A) = \lim_{n\to\infty}\|A^n\|^{1/n}$. One checks by induction that

$$|A^n f| \leq t^{n(p+1)}\frac{\Gamma^n(p+1)}{\Gamma(n(p+1)+1)}\|f\|_{X_0}, \quad n \geq 1, \tag{5.47}$$

where $\|f\|_{X_0}$ is the norm in X_0 and $\Gamma(z)$ is the Gamma function.

The conclusion of Lemma 5.6 follows immediately from (5.47) and from the asymptotic of $\Gamma(p(n+1)+1)$ as $n \to +\infty$.

Lemma 5.10 is proved. □

Lemma 5.11   *Let A be a linear operator in a Banach space Z. If $f = Af + f_0$ and $r(A) = 0$, then $f = \sum_{j=0}^{\infty} A^j f_0$ for any element $f_0 \in Z$. If $f_0 = 0$, then $f = 0$.*

Proof. The conclusion of this Lemma follows from a well-known variant of Lemma 5.11 with the assumption $\|A\| < 1$.

Lemma 5.11 is proved. □

Denote $\Phi_\lambda(t) := \frac{t_+^{\lambda-1}}{\Gamma(\lambda)}$, $\lambda \neq 0, -1, -2, \ldots$, where $t_+^{\lambda-1} := 0$ for $t < 0$ and $t_+^{\lambda-1} := t^{\lambda-1}$ for $t > 0$. This Φ_λ is defined as a distribution, $\Phi_\lambda \star \Phi_\mu = \Phi_{\lambda+\mu}$, $\Phi_\lambda \star \Phi_{-\lambda} = I$, where $\star$ here denotes the convolution in $[0, \infty)$ and I is the identity operator whose kernel is the δ-function.

The convolution of the distribution $\Phi_\lambda(t)$ with any distribution vanishing for $t < 0$ is well defined; see Section 5.10 (see also [4]).

The following lemma allows one to claim that a solution of an integral equation (inequality) with a strongly singular kernel solves an integral equation (inequality) with an integrable kernel.

This result is proved in [20].

Lemma 5.12   If $h = h_0 + \Phi_{-\lambda} \star h$, then $h = \Phi_\lambda \star h - \Phi_\lambda \star h_0$, $\lambda > 0$.

Proof. Apply the operator $\Phi_\lambda \star$ to the equation $h = h_0 + \Phi_{-\lambda} \star h$ and use the property

$$\Phi_\lambda \star \Phi_{-\lambda} = I.$$

This is possible to do as shown in Section 5.10.

Lemma 5.12 is proved.   □

Let us continue with the proof of Theorem 5.7.

One has $\int_0^t (t - s)^{-\frac{5}{4}} \psi \, ds = \Gamma\left(-\frac{1}{4}\right) \Phi_{-\frac{1}{4}} \star \psi$. Inequality (5.45) can be written as

$$\psi \le \psi_0 + c\Gamma\left(-\frac{1}{4}\right) \Phi_{-\frac{1}{4}} \star \psi, \quad \psi_0 := \mathcal{N}\left(|\xi| \tilde{F}\right), \tag{5.48}$$

and $\Gamma\left(-\frac{1}{4}\right) = -4\Gamma(3/4) := -b^{-1}, b > 0$. Applying to (5.48) the operator $\Phi_{1/4} \star$ and multiplying by $c^{-1}b$, one gets

$$\psi \le c^{-1}b\Phi_{1/4} \star \psi_0 - c^{-1}b\Phi_{1/4} \star \psi. \tag{5.49}$$

Using Lemmas 5.10 and 5.11, one derives from (5.49) by iterations that

$$\psi(t) \le h(t), \tag{5.50}$$

where $h(t)$ is the unique solution of the equation

$$h = c^{-1}b\Phi_{1/4} \star \psi_0 - c^{-1}b\Phi_{1/4} \star h. \tag{5.51}$$

Equation (5.51) is solvable by iterations and the iterations converge by Lemma 5.11. The solution h is bounded by a constant depending only on the data, that is, on f, v_0 and $T, 0 \le t \le T$. By inequality (5.50), the function $\psi = \psi(t)$ is bounded by $h = h(t), 0 \le t \le T$. Since $T > 0$ is arbitrary, the second estimate (5.43) is proved. If F decays sufficiently fast as $|x| + t \to \infty$ then $h(t) \to 0$ as $t \to \infty$ and so is $\psi(t)$.

Theorem 5.7 is proved.   □

In the above argument, the notion of convolution of distributions was used: The integral $\int_0^t (t - s)^{-\frac{5}{4}} \psi \, ds$ may diverge classically, but is well defined in the sense of distributions. Convolution of two non-negative distributions with supports on $[0, \infty)$ is a non-negative distribution. The above is discussed in detail at the end of Chapter 5.

A less precise estimate

$$\psi(t) \le c^{-1} b \Phi_{1/4} \star \psi_0 \tag{5.52}$$

than estimate (5.50) can be obtained from Equation (5.49) if one takes into account that $c^{-1} b \Phi_{1/4} \star \psi \ge 0$.

Using formulas (5.41), (5.44), and (5.43) one can estimate $|v(x, t)|$ and $|\tilde{v}(\xi, t)|$.

5.9   PROOF OF THE EXISTENCE OF THE SOLUTION

Since problem (5.40) is equivalent to (5.41) and to (5.44), it is sufficient to prove that (5.44) has a solution and this solution is unique in the space Y.

Proof of the uniqueness of the solution to (5.44) **in the space** Y. Let $\tilde{v}_1$ and $\tilde{v}_2$ solve Equation (5.44). Define

$$w := \tilde{v}_1 - \tilde{v}_2, \quad \mathcal{N}(w) := q, \quad \mathcal{N}(|\xi|w) := Q, \tag{5.53}$$

Subtracting from Equation (5.44) for $\tilde{v}_1$ Equation (5.44) for $\tilde{v}_2$, one gets the equation

$$w = -\int_0^t \tilde{G} \left[w \star (-i\xi \tilde{v}_2) + \tilde{v}_2 \star (-i\xi w) \right] ds, \tag{5.54}$$

where the $\star$ is the convolution in $\mathbb{R}^3$. Taking the norm $\mathcal{N}$ of both sides of (5.54), denoting $z := q + Q$, using estimates (5.43) and formulas (5.46) and (5.53), one derives the following relation:

$$z \le c \left(\Gamma(1/4)\Phi_{1/4} \star z + \Gamma\left(-\frac{1}{4}\right) \Phi_{-\frac{1}{4}} \star z \right), \tag{5.55}$$

where $\star$ here denotes the convolution in $[0, \infty)$. The operator $\Phi_\lambda \star$ with $\lambda > 0$ maps the set of non-negative functions (or distributions) into itself, it preserves inequality sign. Applying the operator $\Phi_{1/4} \star$ to (5.55), taking into account that $\Gamma\left(-\frac{1}{4}\right) = -4\Gamma(3/4)$, $\Phi_\lambda \star \Phi_\mu = \Phi_{\lambda+\mu}$, and $\Phi_\lambda \star \Phi_{-\lambda} = I$, one gets:

$$4c\Gamma(3/4)z \le c\Gamma(1/4)\Phi_{1/2} \star z - \Phi_{1/4} \star z. \tag{5.56}$$

It follows from (5.56) and (5.50) that $z = 0$ (because $f_0 = 0$ in our case; see Lemma 5.12). This proves the uniqueness of the solution to problem (5.40) in the space Y for any $T > 0, 0 \le t \le T$.
□

Proof of the existence of the solution to (5.44). Let us show that the operator B in (5.44) is a contraction in X for a sufficiently small τ, $0 \le t \le \tau$. With $w = \tilde{v}_1 - \tilde{v}_2$, one obtains from (5.44) the following equation:

$$B(\tilde{v}_1) - B(\tilde{v}_2) = -\int_0^t \tilde{G} \left[w \star (-i\xi \tilde{v}_1) + \tilde{v}_2 \star (-i\xi w) \right] ds. \tag{5.57}$$

Here the $\star$ denotes the convolution in $\mathbb{R}^3$. Using the second estimate (5.43), one derives:

$$|\tilde{v}_2 \star (-i\xi w)| \leq \mathcal{N}(w)\mathcal{N}\left((|\xi| + |\xi'|)\,\tilde{v}_2\left(\xi'\right)\right) \leq c\mathcal{N}(w)\left(1 + |\xi|^2\right)^{1/2}. \tag{5.58}$$

Using estimates (5.43), (5.46), and (5.58), one obtains from (5.57) the following inequality:

$$\mathcal{N}\left(B\left(\tilde{v}_1\right) - B\left(\tilde{v}_2\right)\right) \leq c\left(\int_0^t (t-s)^{-\frac{3}{4}} q\,ds + \int_0^t (t-s)^{-\frac{5}{4}} q\,ds\right), \quad q = \mathcal{N}(w). \tag{5.59}$$

Rewrite (5.59) as

$$\mathcal{N}\left(B\left(\tilde{v}_1\right) - B\left(\tilde{v}_2\right)\right) \leq c\left(\Gamma(1/4)\Phi_{1/4} \star q - 4\Gamma(3/4)\Phi_{-\frac{1}{4}} \star q\right), \tag{5.60}$$

where the $\star$ denotes the convolution in $[0, \infty)$. Applying $\lambda_1 \Phi_{1/4}\star$ to (5.60), one gets

$$\lambda_1 \Phi_{\frac{1}{4}} \star \mathcal{N}\left(B\left(\tilde{v}_1\right) - B\left(\tilde{v}_2\right)\right) \leq \lambda_1 c\Gamma(1/4)\Phi_{1/2} \star q - q, \quad \lambda_1^{-1} := 4c\Gamma(3/4) > 0. \tag{5.61}$$

Since $q \geq 0$, one has

$$\lambda_1 \Phi_{\frac{1}{4}} \star \mathcal{N}\left(B\left(\tilde{v}_1\right) - B\left(\tilde{v}_2\right)\right) \leq \lambda_1 c\Gamma(1/4)\Phi_{1/2} \star q. \tag{5.62}$$

From (5.62), one derives

$$\mathcal{N}\left(B\left(\tilde{v}_1\right) - B\left(\tilde{v}_2\right)\right) \leq c\Gamma(1/4)\Phi_{1/4} \star 1 \sup_{t\in[0,\tau]} q(t). \tag{5.63}$$

Clearly, $\lim_{t\to 0} \Phi_{1/4} \star 1 = 0$. Thus, B is a contraction on a ball B_R, $B_R := \{q : q \leq R\}$, if τ is sufficiently small, $0 \leq t \leq \tau$, and $R > 0$ is an arbitrary large fixed number. Consequently, the operator B in (5.44) is a contraction on B_R for $t \leq \tau$. Therefore the solution to (5.44) exists in X for $t \leq \tau$. The *a priori* estimates (5.43) do not depend on $\tau \leq T$. Therefore one can repeat the argument for $\tau \leq t \leq 2\tau$ considering the initial value to be $\tilde{v}(\xi, \tau)$ and the free term to be $\tilde{F}(\xi, t)$, $t \in [\tau, 2\tau]$. The solution $\tilde{v}$ is unique in Y, as we have proved. So, one gets the existence of the solution on $0 \leq t \leq 2\tau$. Continue this process and in finitely many steps get the existence of the unique in X solution in $[0, T]$. Since $T > 0$ is arbitrary, the solution exists for all $T > 0$.

Theorem 5.8 is proved. $\qquad\qquad\qquad\qquad\qquad\qquad\qquad\qquad\qquad\qquad\qquad\square$

It follows from Theorem 5.8 that there cannot be turbulent motions of fluid in the NS problem in the whole space $\mathbb{R}^3$ if the data f and v_0 are smooth and rapidly decaying.

5.10 CONVOLUTION AND POSITIVENESS OF DISTRIBUTIONS

Let us give a detailed discussion of our proof of Lemma 5.12.

A distribution $f \in K'$ is called non-negative if for any test function $\phi \in K, \phi \geq 0$ one has $(f, \phi) \geq 0$. Here (f, ϕ) is the value of the distribution f on a test function ϕ. Very often this value is formally written as $\int f \phi dt$, although the integral may diverge in the classical sense.

The distribution $\Phi_\lambda := \frac{t_+^{\lambda-1}}{\Gamma(\lambda)}$, $\lambda \neq 0, -1, -2, \ldots$ is a non-negative distribution if $\lambda > 0$. Indeed, $\Gamma(\lambda) \geq 0$ if $\lambda \geq$) and $t_+^\lambda \geq 0$ if $\lambda > 0$. The support of Φ_λ is $\mathbb{R}_+ = [0, \infty)$.

The distribution $\psi = \psi(t) := \mathcal{N}(\nabla v)$ is non-negative and its support is $[0, \infty)$ because $v = 0$ for $t < 0$.

The convolution $\Phi_\lambda \star \psi$ is well defined for distributions with supports belonging to $[0, \infty)$. This definition is well known [4]. Namely, if $supp f \subset [0, \infty)$ and $supp \psi \subset [0, \infty)$, then

$$(f \star \psi, \phi) = \int_0^\infty \int_0^\infty f(t-s)\psi(s)\phi(t) \, ds \, dt$$
$$= \int_0^\infty \int_0^\infty f(\xi)\psi(s)\phi(\xi+s) \, ds \, d\xi. \tag{5.64}$$

Since the test function ϕ has compact support, it follows that the function

$$\eta(\xi) := \int_0^\infty \psi(s)\phi(s+\xi) ds$$

is infinitely differentiable and vanishes for sufficiently large ξ.

Consequently, $supp \eta \subset (-\infty, c)$, where $c > 0$ is some constant. Therefore the quantity

$$(f, \eta) = \int_0^\infty f(\xi)\eta(\xi)d\xi \tag{5.65}$$

makes sense. Indeed, the distribution f has support in $[0, \infty)$, while the test function η has support bounded from above. Therefore the intersection of these supports is a bounded set. That is why the definitions (5.64) and (5.65) make sense.

In particular, the convolution of Φ_λ and ψ makes sense although for $\lambda = -\frac{1}{4}$ the integral $\int_0^t (t-s)^{-5/4}\psi(s)ds$ may diverge in the classical sense.

The distributions $\Gamma\left(-\frac{1}{4}\right)\Phi_{-\frac{1}{4}}$ and ψ are non-negative. Their convolution is also a non-negative distribution.

A convolution of a non-negative distribution with a non-negative distribution $\Phi_{\frac{1}{4}}$ is a non-negative distribution.

Therefore, taking a convolution of $\Phi_{\frac{1}{4}}$ with the inequality $0 \leq g$, where g is a distribution, one gets $0 \leq \Phi_{\frac{1}{4}} \star g$.

CHAPTER 6

Inverse Problem of Potential Theory

6.1 STATEMENT OF THE PROBLEM

In this chapter, the author's paper [52] is used and the presentation follows closely the one in this paper. This chapter can be read independently of the previous chapters.

Suppose there are two bodies D_j, $j = 1, 2$, uniformly charged with charge density 1. Let the corresponding potentials be $u_j(x) = \int_{D_j} \frac{dy}{4\pi|x-y|}$. Assume that $u_1(x) = u_2(x)$ for $|x| > R$, where $R > 0$ is a large number.

The inverse problem of potential theory is:
Does this imply that $D_1 = D_2$?

P. Novikov in 1938 (see [15]) proved a uniqueness theorem for the solution of inverse problem (IP) of potential theory under a special assumption; see Proposition 6.1 below.

Let

$$u(x) = \int_D g_0(x, y)dy, \quad g_0(x, y) := \frac{1}{4\pi|x - y|}, \tag{6.1}$$

where $D \subset \mathbb{R}^3$ is a bounded, connected, C^2−smooth domain.

We use the following notations: D_j, $j = 1, 2$, are two different domains D, S_j is the boundary of D_j, $D'_j = \mathbb{R}^3 \setminus D_j$, S^2 is the unit sphere in $\mathbb{R}^3$, B_R is the ball of radius R, centered at the origin, $B'_R = \mathbb{R}^3 \setminus B_R$, $D_j \subset B_R$, $D_{12} := D_1 \cup D_2$, $D'_{12} := \mathbb{R}^3 \setminus D_{12}$, $S_{12} := \partial D_{12}$, $S^{12} := \partial D^{12}$, $\mathcal{D} := D^{12} = D_1 \cap D_2$, $\mathcal{D}' = \mathbb{R}^3 \setminus \mathcal{D}$, $\mathcal{S}$ is the boundary of $\mathcal{D}$, $|D|$ is the volume of D and $u_j(x) = \int_{D_j} g_0(x, y)dy$, $j = 1, 2$.

Proposition 6.1 *If $u_1(x) = u_2(x)$ for $|x| > R$, then $D_1 = D_2$ provided that D_j, $j = 1, 2$, are star-shaped with respect to a common point.*

In [13] this result is generalized: The existence of the common point with respect to which D_1 and D_2 are star-shaped is not assumed, but D_j are still assumed star-shaped.

In [26, p. 334], a new proof of Proposition 6.1 was given.

The goal of this chapter is to give a new method for a proof of an important generalization of Proposition 6.1.

In this generalization the novel points are:

(a) the assumptions about star-shapeness of D_j, $j = 1, 2$, are discarded (see Theorem 6.6 below);

(b) a new approach to the IP is developed;

and

(c) a similar inverse problem is studied in the case when $g_0(x, y)$ is replaced by the Green's function of the Helmholtz operator, $g(x, y) := \frac{e^{ik|x-y|}}{4\pi|x-y|}$, $k = const > 0$ is fixed, and u_j is replaced by

$$U_j(x) = \int_{D_j} g(x, y) dy.$$

The result is formulated in Theorem 6.7, below.

The idea of our proof does not use the basic idea of [15], [13], or [26].

Our proof is based on some lemmas.

Lemma 6.2   *If $R(\phi)$ is the element of the rotation group in $\mathbb{R}^3$, then $\frac{\partial R(\phi)x}{\partial\phi}|_{\phi=0} = [\alpha, x]$, where $[x, y]$ is the cross product of two vectors, α is the unit vector around which the rotation by the angle ϕ takes place, and x is an arbitrary vector.*

This lemma is proved in [26, p. 416].

Lemma 6.3   *The set of restrictions on S of all harmonic in B_R functions is dense in $L^2(S)$, where S is the boundary of D.*

Proof of Lemma 6.3. Let us assume the contrary and derive a contradiction. Without loss of generality, one may assume f to be real-valued. Suppose $f \neq 0$ is orthogonal in $L^2(S)$ to any harmonic function h, that is,

$$\int_S fh ds = 0 \tag{6.2}$$

for all harmonic functions in B_R.

Define

$$v(x) := \int_S g_0(x, s) f ds.$$

Assumption (6.2) implies $v(x) = 0$ in $\mathbb{R}^3$. Indeed, there exists a unique solution to the problem

$$\Delta h = 0 \quad \text{in} \quad D, \quad h|_S = f.$$

For this h, Equation (6.2) implies that $f = 0$, so $v = 0$ in $\mathbb{R}^3$.

If S is a union of two smooth closed surfaces, then our argument remains valid.

For example, if S_1 intersects S_2, then there exists a unique harmonic function in the region $D_{12} \setminus D^{12}$ which takes prescribed values on S_{12} and on S^{12}.

Here $D_{12} = D_1 \cup D_2$, S_{12} is the boundary of D_{12}, $D^{12} = D_1 \cap D_2$, and S^{12} is the boundary of D^{12}.

Lemma 6.3 is proved, see also Lemma 4.2 in Chapter 4.   □

Remark 6.4   The proof of Lemma 6.3 is valid for closed surfaces S which are not necessarily connected. For example, it is valid for S, which is a union of two surfaces which intersects.

It is known (see, for example, [26]) that

$$g_0(x, y) = g_0(|x|) \sum_{\ell \geq 0} \frac{|y|^\ell}{|x|^\ell} \overline{Y_\ell(y^0)} Y_\ell(x^0), \quad |x| > |y|, \tag{6.3}$$

where $x^0 := x/|x|$, Y_ℓ are the spherical harmonics, normalized in $L^2(S^2)$; the overline stands for the complex conjugate and $|y|^\ell Y_\ell(y^0)$ are harmonic functions. The set of these functions for all $\ell \geq 0$ is dense in the set of all harmonic functions in B_R.

Lemma 6.5   *If S is a smooth closed surface and $[s, N] = 0$ on S, then S is a sphere.*

Proof of Lemma 6.5.   Let $s = s(p, q)$ be a parametric equation of S. Then N is proportional to $[s_p, s_q]$, where s_p is the partial derivative $\frac{\partial s}{\partial p}$. If $[s, N] = 0$, then $[s, [s_p, s_q]] = 0$. Thus, $s_p s \cdot s_q - s_q s \cdot s_p = 0$, where $s \cdot s_q$ is the dot product of two vectors. Since S is smooth, vectors s_p and s_q are linearly independent on S. Therefore, $\frac{\partial s \cdot s}{\partial p} = 0$, and $\frac{\partial s \cdot s}{\partial q} = 0$. Consequently, $s \cdot s = const$. This means that S is a sphere.

Lemma 6.5 is proved.   □

Lemma 6.5 is Lemma 11.2.2 in [26], see also Theorem 3.7 in Chapter 3. Its short proof is included for convenience of the reader.

It follows from our proof that if S has finitely many points of non-smoothness, then the parts of S, joining these points, are spherical segments.

Our new results are the following theorems.

Theorem 6.6   *If $u_1(x) = u_2(x)$ for $|x| > R$, then $D_1 = D_2$.*

Theorem 6.7   *There may exist, in general, countably many different D_j such that the corresponding potentials U_j are equal in B'_R for sufficiently large $R > 0$.*

Remark 6.8   If

$$V_j := \int_{S_j} g(x, t) dt, \quad j = 1, 2,$$

then there exist $S_1 \neq S_2$ for which $V_1(x) = V_2(x) = 0 \; \forall x \in B'_R$.

An example can be constructed similarly to the one given in the proof of Theorem 6.7. In Section 6.2 proofs are given.

6.2   PROOFS

Proof of Theorem 6.6. If $u_1 = u_2$ for all $x \in B'_R$, then it follows from the asymptotic of u_j as $|x| \to \infty$ that $|D_1| = |D_2|$. Thus, the case $D_1 \subset D_2$ is not possible if $u_1 = u_2$ for all $x \in B'_R$.

The functions u_j are harmonic functions in D'_j, that is, $\Delta u_j = 0$ in D'_j. If $u_1 = u_2$ in B'_R and $D_1 \neq D_2$, then $u_1 = u_2$ in D'_{12} by the unique continuation property for harmonic functions. Let $w := u_1 - u_2$. Then $w = 0$ in D'_{12},

$$\Delta w = \chi_2 - \chi_1, \tag{6.4}$$

where χ_j is the characteristic function of D_j.

Let h be an arbitrary harmonic function in B_R. Then

$$\int_{D_2} h(x)dx - \int_{D_1} h(x)dx = 0,$$

as one gets multiplying (6.4) by h, integrating by parts and taking into account that w vanishes outside D_{12}.

If $h(x)$ is harmonic, so is $h(R(\phi)x)$. Thus,

$$\int_{D_2} h(R(\phi)x)dx - \int_{D_1} h(R(\phi)x)dx = 0. \tag{6.5}$$

Differentiate (6.5) with respect to ϕ and then set $\phi = 0$. Using Lemma 6.2, one gets:

$$\int_{D_2} \nabla h \cdot [\alpha, x]dx - \int_{D_1} \nabla h \cdot [\alpha, x]dx = 0, \tag{6.6}$$

where $\alpha \in S^2$ is arbitrary, and h is an arbitrary harmonic function in B_R. Since $\nabla h \cdot [\alpha, x] = \nabla \cdot (h[\alpha, x])$, it follows from (6.6) and the divergence theorem that

$$\int_{S_2} Nh[\alpha, s]ds - \int_{S_1} Nh[\alpha, s]ds = 0, \tag{6.7}$$

for all $\alpha \in S^2$ and all harmonic h in B_R. Here N is the unit normal to the boundary pointing out of D_j.

If $D_1 = D_2$, then (6.7) is an identity. Suppose that $D_1 \neq D_2$. Since α is arbitrary, it follows from (6.7) that

$$\int_{S_2} h[N, s]ds - \int_{S_1} h[N, s]ds = 0, \tag{6.8}$$

for all harmonic in B_R functions h.

By Lemma 6.3 and Remark 6.4, it follows from (6.8) that $[N, s] = 0$ on S_2 and on S_1.

By Lemma 6.5, it follows that S_1 and S_2 are spheres, so D_1 and D_2 are balls. These balls must be of the same radius, as was mentioned earlier.

Now we have a contradiction unless $D_1 = D_2$, because two balls with the same constant density of the charge and with the same total charge but with different centers cannot have the same potential in B'_R. This follows from the explicit formula for their potentials. If these balls have the same center and the same radius, then $D_1 = D_2$.

Theorem 6.6 is proved. □

Proof of Theorem 6.7. Let $D_j = B_{a_j}$, where $a_j > 0$ are some numbers which are chosen below. Then $U_j(x) := \int_{|y| \leq a_j} g(x, y) dy = 0$ in the region B'_{a_j} if and only if

$$\int_0^{a_j} r^2 j_0(kr) dr = 0,$$

where $j_0(r)$ is the spherical Bessel function,

$$j_0(kr) := \left(\frac{\pi}{2kr}\right)^{1/2} J_{1/2}(kr) = \frac{\sin(kr)}{kr}.$$

This follows from the formula $U_j(x) = \int_{B_{a_j}} g(x, y) dy$, and from the known formula for g(x,y) (see, for example, [26]):

$$g(x, y) = \sum_{\ell \geq 0} ik j_\ell(k|y|) h_\ell(k|x|) \overline{Y_\ell(y^0)} Y_\ell(x^0), \quad |y| < |x|.$$

Here Y_ℓ are the normalized spherical harmonics (see [26, p. 261]), j_ℓ and h_ℓ are the spherical Bessel and Hankel functions (see [26, p. 262]), and the known formula

$$\int_{S^2} Y_\ell\left(y^0\right) dy^0 = 0, \qquad \ell > 0$$

was used.

One has

$$\int_0^{a_j} r^2 j_0(kr) dr = \frac{\sin(ka_j)}{k^2} - \frac{a_j \cos(ka_j)}{k^2} = 0$$

if and only if

$$\tan\left(ka_j\right) = a_j.$$

This equation has countably many positive solutions. To each of these solutions there corresponds a ball B_{a_j} such that $U_j = 0$ in B'_{a_j}. Thus, there are many different balls for which U_j are the same in B'_R, namely $U_j = 0$ in B'_R for $R > a_j$. Theorem 6.7 is proved. □

Bibliography

[1] Aharonov, D., Schiffer, M., and Zalcman, L., Potato kugel, *Israel Journal of Mathematics*, 40, N3-4, pp. 331–339, 1981. DOI: 10.1007/bf02761373. 3

[2] Bateman, H. and Erdelyi, A., *Tables of Integral Transforms*, McGraw-Hill, New York, 1954. 47

[3] Chakalov, L., Sur un probléme de D. Pompeiu, *Godishnik University of Sofia, Faculty of Physics-Mathematics*, 40, pp. 1–14, 1944. 26

[4] Gel'fand, I. and Shilov, G., *Generalized Functions*, vol. 1, AMS Chelsea Publishing, 1964. 53, 57

[5] Gilbarg, D. and Trudinger, N., *Elliptic Partial Differential Equations of Second Order*, Springer-Verlag, New York, 1983. DOI: 10.1007/978-3-642-61798-0.

[6] Hoang, N. S. and Ramm, A. G., Symmetry problems 2, *Annales Polonici Mathematici*, 96, N1, pp. 61–64, 2009. DOI: 10.4064/ap96-1-5. 3, 29

[7] Ladyzhenskaya, O. A., *The Mathematical Theory of Viscous Incompressible Flow*, Gordon and Breach, New York, 1969. DOI: 10.1063/1.3051412. 39, 45, 49

[8] Landau, L. and Lifshitz, E., *Fluid Mechanics*, Pergamon Press, Oxford, 1984. 39

[9] Landau, L. and Lifshitz, E., *Quantum Mechanics*, Pergamon Press, Oxford, 1984. DOI: 10.1016/C2013-0-02793-4. 8

[10] Landau, L. and Lifshitz, E., *Electrodynamics of Continuous Media*, Pergamon Press, Oxford, 1984.

[11] Lebedev, N., *Special Functions and their Applications*, Dover, New York, 1972. 22, 39

[12] Lemarie-Rieusset, P., *The Navier-Stokes Problem in the 21st Century*, CRC Press, Boca Raton, FL, 2016. DOI: 10.1201/9781315373393. 4, 39

[13] Margulis, A., Equivalence and uniqueness in the inverse problem of potential theory for homogeneous star-shaped bodies, *Doklady of the USSR Academy of Science*, 312, pp. 577–580, 1990. 59, 60

[14] Mikhlin, S. and Prössdorf, S., *Singular Integral Equations*, Springer-Verlag, Berlin, 1986. DOI: 10.1016/b978-0-08-010852-0.50011-6. 44

[15] Novikov, P., On the uniqueness of inverse problem of potential theory, *Doklady of the USSR Academy of Science*, 19, pp. 165–168, 1938. 4, 59, 60

[16] Pompeiu, D., Sur une propriety integrale des functiones de deux variables reelles, *Bulletin of Science, Royal Academy of Belgique*, 5, N15, pp. 265–269, 1929. 26

[17] Ramm, A. G., Large-time behavior of the weak solution to 3D Navier-Stokes equations, *Applied Mathematics Letters*, 26, pp. 252–257, 2013. DOI: 10.1016/j.aml.2012.09.003. 46

[18] Ramm, A. G., Existence and uniqueness of the global solution to the Navier-Stokes equations, *Applied Mathematics Letters*, 49, pp. 7–11, 2015. DOI: 10.1016/j.aml.2015.04.008. 39

[19] Ramm, A. G., Large-time behavior of solutions to evolution equations, *Handbook of Applications of Chaos Theory*, C. Skiadas, Ed., pp. 183–200, Chapman and Hall/CRC, 2016. DOI: 10.1201/b20232. 46

[20] Ramm, A. G., Existence of the solutions to convolution equations with distributional Kernels, *Global Journal of Mathematics Analysis*, 6(1), pp. 1–2, 2018. 54

[21] Ramm, A. G., Necessary and sufficient condition for a scattering amplitude to correspond to a spherically symmetric scatterer, *Applied Mathematics Letters*, 2, pp. 263–265, 1989. DOI: 10.1016/0893-9659(89)90066-9. 5

[22] Ramm, A. G., Symmetry properties for scattering amplitudes and applications to inverse problems, *Journal of Mathematical Analysis and Applications*, 156, pp. 333–340, 1991. DOI: 10.1016/0022-247x(91)90401-k. 5

[23] Ramm, A. G. and Weaver, O., Necessary and sufficient condition for the potential to be spherically symmetric, *Inverse Problems*, 5, pp. L45–47, 1989. DOI: 10.1088/0266-5611/5/5/001. 5

[24] Ramm, A. G., Minimization of the total radiation from an obstacle by a control function on a part of the boundary, *Journal of Inverse and Ill-posed Problem*, 4, N6, pp. 531–534, 1996. DOI: 10.1515/jiip.1996.4.6.531. 34

[25] Ramm, A. G., *Wave Scattering by Small Bodies of Arbitrary Shapes*, World Scientific Publishers, Singapore, 2005. DOI: 10.1142/9789812701206.

[26] Ramm, A. G., *Inverse Problems*, Springer, New York, 2005. 3, 17, 27, 59, 60, 61, 63

[27] Ramm, A. G., The Pompeiu problem, *Applicable Analysis*, 64, N1-2, pp. 19–26, 1997. DOI: 10.1080/00036819708840520. 3

[28] Ramm, A. G., A symmetry problem, *Annales Polonici Mathematici*, 92, pp. 49–54, 2007. DOI: 10.4064/ap92-1-5. 3, 29, 31

[29] Ramm, A. G., Invisible obstacles, *Annales Polonici Mathematici*, 90, N2, pp. 145–148, 2007. DOI: 10.4064/ap90-2-4. xiii, 3, 4, 29, 34

[30] Ramm, A. G., Symmetry problem, *Proc. of the American Mathematical Society*, 141, N2, pp. 515–521, 2013. DOI: 10.1090/s0002-9939-2012-11400-5. 3, 23

[31] Ramm, A. G., The Pompeiu problem, *Global Journal of Mathematical Analysis (GJMA)*, 1, N1, pp. 1–10, 2013. http://www.sciencepubco.com/index.php/GJMA/issue/curr ent DOI: 10.14419/gjma.v1i1.728. 3, 17, 26

[32] Ramm, A. G., A symmetry result for strictly convex domains, *Analysis*, 35(1), pp. 29–32, 2015. DOI: 10.1515/anly-2014-1257.

[33] Ramm, A. G., Solution to the Pompeiu problem and the related symmetry problem, *Applied Mathematics Letters*, 63, pp. 28–33, 2017. DOI: 10.1016/j.aml.2016.07.015. 3, 17, 26

[34] Ramm, A. G., Global existence, uniqueness and estimates of the solution to the Navier-Stokes equations, *Applied Mathematics Letters*, 74, pp. 154–160, 2017. DOI: 10.1016/j.aml.2017.05.009. 39

[35] Ramm, A. G., Solution of the Navier-Stokes problem, *Applied Mathematics Letters*, 87, pp. 160–164, 2019. DOI: 10.1016/j.aml.2018.07.034. 39

[36] Ramm, A. G., Symmetry problem 1, *Journal of Advances in Mathematics (JAM)*, 15, pp. 1–4, 2018. DOI: 10.1090/s0002-9939-2012-11400-5. 17

[37] Ramm, A. G., *Scattering of Acoustic and Electromagnetic Waves by Small Bodies of Arbitrary Shapes. Applications to Creating New Engineered Materials*, Momentum Press, New York, 2013. DOI: 10.5643/9781606506226. xiii, 4

[38] Ramm, A. G., Creating materials with a desired refraction coefficient, *IOP Concise Physics*, Morgan & Claypool Publishers, San Rafael, CA, 2017. xiii, 4

[39] Ramm, A. G., Finding a method for producing small impedance particles with prescribed boundary impedance is important, *Journal of Physics Research and Applications*, 1:1, pp. 1–3, 2017.

[40] Ramm, A. G., Many-body wave scattering problems for small scatterers and creating materials with a desired refraction coefficient, in the book *Mathematical Analysis and Applications: Selected Topics*, M. Ruzhansky, H. Dutta, and R. Agarwal, Eds., Chapter 3, pp. 57–76, Wiley, Hoboken, NJ, 2018. DOI: 10.1002/9781119414421. xiii

[41] Ramm, A. G., Scattering of EM waves by many small perfectly conducting or impedance bodies, *Journal of Mathematics and Physics (JMP)*, 56, N9, 091901, 2015. DOI: 10.1063/1.4929965.

[42] Ramm, A. G., EM wave scattering by many small impedance particles and applications to materials science, *The Open Optics Journal*, 9, pp. 14–17, 2015. `http://benthamopen.co m/TOOPTSJ/VOLUME/9/` DOI: 10.2174/1874328501509010014.

[43] Ramm, A. G., *Scattering by Obstacles*, D. Reidel Publishing Company, Dordrecht, Holland, 1986. DOI: 10.1007/978-94-009-4544-9. 35

[44] Ramm, A. G., Uniqueness theorem for inverse scattering problem with non-over-determined data, *Journal of Physics A*, FTC, 43, 112001, 2010. DOI: 10.1088/1751-8113/43/11/112001. 4

[45] Ramm, A. G., Uniqueness of the solution to inverse scattering problem with backscattering data, *Eurasian Mathematical Journal (EMJ)*, 1, N3, pp. 97–111, 2010. DOI: 10.1063/1.3666985. xiii, 4

[46] Ramm, A. G., Uniqueness of the solution to inverse scattering problem with scattering data at a fixed direction of the incident wave, *Journal of Mathematics and Physics*, 52, 123506, 2011. DOI: 10.1063/1.3666985. xiii, 4

[47] Ramm, A. G., *Dynamical Systems Method for Solving Operator Equations*, Elsevier, Amsterdam, 2007. DOI: 10.1016/s1007-5704(03)00006-6. 4

[48] Ramm, A. G. and Hoang, N. S., *Dynamical Systems Method and Applications. Theoretical Developments and Numerical Examples*, Wiley, Hoboken, NJ, 2012. DOI: 10.1002/9781118199619. 4

[49] Ramm, A. G., *Scattering by Obstacles and Potentials*, World Scientific Publishers, Singapore, 2017. DOI: 10.1142/10473. xiii, 1, 2, 4, 5, 6, 7, 8, 11, 12, 13, 32, 35, 36

[50] Ramm, A. G., Necessary and sufficient condition for a surface to be a sphere, *Open Journal of Mathematical Analysis (OMA)*, 2, issue 2, pp. 51–52, 2018.

[51] Ramm, A. G., Old symmetry problem revisited, *Open Journal of Mathematical Analysis, (OMA)*, 2, N2, pp. 89–92, 2018. 15, 23

[52] Ramm, A. G., Inverse problem of potential theory, *Applied Mathematics Letters*, 77, pp. 1–5, 2018. DOI: 10.1016/j.aml.2017.09.012. 59

[53] Ramm, A. G., On the Navier-Stokes problem, *Journal of Advances in Mathematics*, 16, pp. 1–5, 2019. 39

[54] Ramm, A. G., Inverse scattering with non-over-determined data, *Journal of Advances in Mathematics*, 16, pp. 1–4, 2019. DOI: 10.1109/mmet.2016.7544097. xiii, 4

[55] Ramm, A. G., Symmetry problems for the Helmholtz equation, (submitted). 15

[56] Serrin, J., A symmetry problem in potential theory, *Archive for Rational Mechanics and Analysis*, 43, N4, pp. 304–318, 1971. DOI: 10.1007/bf00250468. 3, 23

Author's Biography

ALEXANDER G. RAMM

Alexander G. Ramm, Ph.D., was born in Russia, immigrated to the U.S. in 1979, and is a U.S. citizen. He is Professor of Mathematics with broad interests in analysis, scattering theory, inverse problems, theoretical physics, engineering, signal estimation, tomography, theoretical numerical analysis, and applied mathematics. He is an author of 690 research papers, 16 monographs, and an editor of 3 books. He has lectured in many universities throughout the world, presented approximately 150 invited and plenary talks at various conferences, and has supervised 11 Ph.D. students. He was Fulbright Research Professor in Israel and in Ukraine, distinguished visiting professor in Mexico and Egypt, Mercator professor, invited plenary speaker at the 7th PACOM, won the Khwarizmi international award, and received other honors. Recently he solved inverse scattering problems with non-over-determined data and the many-body wave-scattering problem when the scatterers are small particles of an arbitrary shape; Dr. Ramm used this theory to give a recipe for creating materials with a desired refraction coefficient, gave a solution to the refined Pompeiu problem and proved the refined Schiffer's conjecture.